FRONTIERS OF ENGINEERING

REPORTS ON LEADING-EDGE ENGINEERING FROM THE 2005 SYMPOSIUM

NATIONAL ACADEMY OF ENGINEERING
OF THE NATIONAL ACADEMIES

THE NATIONAL ACADEMIES PRESS
Washington, D.C.
www.nap.edu

THE NATIONAL ACADEMIES PRESS • 500 Fifth Street, N.W. • Washington, D.C. 20001

NOTICE: This publication has been reviewed according to procedures approved by a National Academy of Engineering report review process. Publication of signed work signifies that it is judged a competent and useful contribution worthy of public consideration, but it does not imply endorsement of conclusions or recommendations by the NAE. The interpretations and conclusions in such publications are those of the authors and do not purport to represent the views of the council, officers, or staff of the National Academy of Engineering.

Funding for the activity that led to this publication was provided by the Air Force Office of Scientific Research, Defense Advanced Research Projects Agency, Department of Defense–DDR&E-Research, National Science Foundation, Department of Homeland Security, General Electric, Microsoft Corporation, Cummins, Inc., John A. Armstrong, and other individual donors.

International Standard Book Number 0-309-10102-6

Additional copies of this report are available from the National Academies Press, 500 Fifth Street, N.W., Lockbox 285, Washington, DC 20001; (800) 624-6242 or (202) 334-3313 (in the Washington metropolitan area); Internet, http://www.nap.edu.

Printed in the United States of America

THE NATIONAL ACADEMIES

Advisers to the Nation on Science, Engineering, and Medicine

The **National Academy of Sciences** is a private, nonprofit, self-perpetuating society of distinguished scholars engaged in scientific and engineering research, dedicated to the furtherance of science and technology and to their use for the general welfare. Upon the authority of the charter granted to it by the Congress in 1863, the Academy has a mandate that requires it to advise the federal government on scientific and technical matters. Dr. Ralph J. Cicerone is president of the National Academy of Sciences.

The **National Academy of Engineering** was established in 1964, under the charter of the National Academy of Sciences, as a parallel organization of outstanding engineers. It is autonomous in its administration and in the selection of its members, sharing with the National Academy of Sciences the responsibility for advising the federal government. The National Academy of Engineering also sponsors engineering programs aimed at meeting national needs, encourages education and research, and recognizes the superior achievements of engineers. Dr. Wm. A. Wulf is president of the National Academy of Engineering.

The **Institute of Medicine** was established in 1970 by the National Academy of Sciences to secure the services of eminent members of appropriate professions in the examination of policy matters pertaining to the health of the public. The Institute acts under the responsibility given to the National Academy of Sciences by its congressional charter to be an adviser to the federal government and, upon its own initiative, to identify issues of medical care, research, and education. Dr. Harvey V. Fineberg is president of the Institute of Medicine.

The **National Research Council** was organized by the National Academy of Sciences in 1916 to associate the broad community of science and technology with the Academy's purposes of furthering knowledge and advising the federal government. Functioning in accordance with general policies determined by the Academy, the Council has become the principal operating agency of both the National Academy of Sciences and the National Academy of Engineering in providing services to the government, the public, and the scientific and engineering communities. The Council is administered jointly by both Academies and the Institute of Medicine. Dr. Ralph J. Cicerone and Dr. Wm. A. Wulf are chair and vice chair, respectively, of the National Research Council.

www.national-academies.org

ORGANIZING COMMITTEE

PABLO G. DEBENEDETTI (Chair), Class of 1950 Professor, Department of Chemical Engineering, Princeton University
LUIS A. NUNES AMARAL, Associate Professor, Department of Chemical and Biological Engineering, Northwestern University
ALLAN J. CONNOLLY, General Manager, Power Generation Systems Engineering, GE Energy
STEPHEN S. INTILLE, Technology Director, House_n Consortium, Department of Architecture, Massachusetts Institute of Technology
KELVIN H. LEE, Associate Professor, School of Chemical and Biomolecular Engineering, Cornell University
GARRICK E. LOUIS, Associate Professor, Department of Systems and Information Engineering, University of Virginia
VISVANATHAN RAMESH, Department Head, Real-Time Vision and Modeling, Siemens Corporate Research, Inc.
AMY SMITH, Instructor, Edgerton Center, Massachusetts Institute of Technology
JOHN M. VOHS, Carl V. Patterson Professor and Chair, Department of Chemical and Biomolecular Engineering, University of Pennsylvania

Staff

JANET R. HUNZIKER, Senior Program Officer
VIRGINIA R. BACON, Senior Program Assistant

Preface

In 1995, the National Academy of Engineering (NAE) initiated the Frontiers of Engineering Program, which brings together about 100 young engineering leaders for annual symposia to learn about cutting-edge research and technical work in different fields of engineering. On September 22–24, 2005, NAE held its eleventh U.S. Frontiers of Engineering Symposium at GE Global Research Center in Niskayuna, New York. Speakers were asked to prepare extended summaries of their presentations, which are reprinted here. The intent of this volume, and of the volumes that preceded it in the series, is to convey the excitement of this unique meeting and to highlight cutting-edge developments in engineering research.

GOALS OF THE FRONTIERS OF ENGINEERING PROGRAM

The practice of engineering is continually changing. Engineers today must be able not only to thrive in an environment of rapid technological change and globalization, but also to work on interdisciplinary teams. Cutting-edge research is being done at the intersections of engineering disciplines, and successful researchers and practitioners must be aware of developments and challenges in areas other than their own.

At the three-day U.S. Frontiers of Engineering Symposium, 100 of this country's best and brightest engineers, ages 30 to 45, have an opportunity to learn from their peers about pioneering work being done in many areas of engineering. The symposium gives engineers from a variety of institutions in academia, industry, and government, and from many different engineering disciplines, an opportunity to make contacts with and learn from individuals whom they would not meet in the usual round of professional meetings. This networking may lead to collaborative work and facilitate the transfer of new techniques and approaches. It is hoped that the exchange of information on current developments in many fields of engineering will lead to insights that may be applicable in specific disciplines

The number of participants at each meeting is limited to 100 to maximize opportunities for interactions and exchanges among the attendees, who are chosen through a competitive nomination and selection process. The choice of topics and speakers for each meeting is made by an organizing committee composed of engineers in the same 30- to 45-year-old cohort as the participants. Each year different topics are covered, and, with a few exceptions, different individuals participate.

Speakers describe the challenges they face and communicate the excitement of their work to a technically sophisticated but nonspecialized audience. Each speaker provides a brief overview of his/her field of inquiry; defines the frontiers of that field; describes experiments, prototypes, and design studies that have been completed or are in progress, as well as new tools and methodologies, and limitations and controversies; and summarizes the long-term significance of his/her work.

THE 2005 SYMPOSIUM

The four general topics for the 2005 meeting were: ID and verification technologies, engineering for developing communities, the engineering of complex systems, and energy resources for the future (see Appendix C). The session on ID and verification technologies was based on the proliferation of cheap and novel sensors, faster computers, and intelligent algorithms that have improved monitoring capabilities and made it possible to identify and track objects and people. Two of the talks in this session were on face recognition systems—an overview highlighting the difficulties and a presentation on challenge problems and independent evaluations in automatic face recognition. A third talk focused on advances in RFID and activity recognition and their potential for improving the quality of life for elderly people.

The second session, engineering for developing communities, addressed challenges and opportunities for engineering to alleviate poverty and promote sustainability. Presentations included: a description of the DISACARE wheelchair project in Zambia, an example of how technologies can be adapted to address local conditions; the impact of engineering advances on the safe water system program of the Centers for Disease Control and Prevention; a discussion of sustainable development through green engineering; and a paper on solar electricity markets in developing nations to illustrate the value of making sustainability science a guiding scientific principle.

What do metabolic pathways and ecosystems, the Internet, and the propagation of HIV infection have in common? The session on the engineering of complex systems suggested answers to this question. The presentations provided an overview of theoretical and experimental tools that are enabling us to tackle more systematically the challenges posed by complex systems. Speakers described the emergence of network theory, one of the most visible components of the body of knowledge that can be applied to the description, analysis, and understanding of complex systems; the engineering of biological systems; and agent-based modeling, which can be used to study a wide variety of systems, from ant colonies, trader behavior in financial systems, and traffic patterns to urban growth and the spread of epidemics. An additional paper on a mathematical formulation of language evolution, which was on the original program but was not presented at the meeting, is included in this volume.

The final session, on energy resources for the future, included presentations on organic-based solar cells, research on hydrogen production and storage sponsored by the U.S. Department of Energy, and the future of fuel cells.

Every year, a distinguished engineer addresses the participants at dinner on the first evening of the symposium. The speaker this year, Shirley Ann Jackson, president of Rensselaer Polytechnic Institute, delivered an inspirational talk entitled, "Engineering for a New World." She linked the critical issues facing us with the importance of nurturing and encouraging the capacity for innovation in future engineers and scientists. The full text of Dr. Jackson's remarks is included in this volume.

NAE is deeply grateful to the following organizations for their support of the 2005 U.S. Frontiers of Engineering Symposium: GE, Air Force Office of Scientific Research, Defense Advanced Research Projects Agency, U.S. Department of Defense-DDR&E Research, U.S. Department of Homeland Security, National Science Foundation, Microsoft Corporation, Cummins Inc., and Dr. John A. Armstrong and other individual donors. NAE would also like to thank the members of the Symposium Organizing Committee (p. iv), chaired by Professor Pablo Debenedetti, for planning and organizing the event.

Contents

DINNER SPEECH

APPENDIXES

ID and Verification Technologies

Introduction

STEPHEN S. INTILLE
Massachusetts Institute of Technology
Cambridge, Massachusetts

VISVANATHAN RAMESH
Siemens Corporate Research, Inc.
Princeton, New Jersey

The profileration of cheaper, novel sensors, faster computers, and intelligent algorithms has made the effective monitoring, identifying, and tracking of objects and persons much more feasible. Modalities for identifying and tracking objects include: radio-frequency identification devices (RFIDs), bar codes, and so on. Modalities for the biometric identification of people include facial-recognition systems, fingerprint analysis, hand-geometry analysis, iris recognition, and many others. The combination of biometrics and digital passports will enable the tracking of individuals and their whereabouts. Moreover, video surveillance systems, which are being increasingly used in public areas, will be able to detect, track, capture, and log faces of individuals and, potentially, match them to faces in databases.

The presentations in this session will provide an overview of the trends and research challenges in technologies in this field. As the capability to identify, track, and monitor individuals improves, we will need safeguards against the misuse of these capabilities. Thus, privacy issues related to the deployment of these technologies will also be discussed.

The presentations will focus on the following topics: (1) an overview of research challenges and face-recognition technology by Peter Belhumeur; (2) an evaluation of biometrics for face recognition by Jonathon Phillips; and (3) the practical use of RFIDs for activity recognition by Matthai Philipose.

Ongoing Challenges in Face Recognition

Peter N. Belhumeur
Columbia University
New York, New York

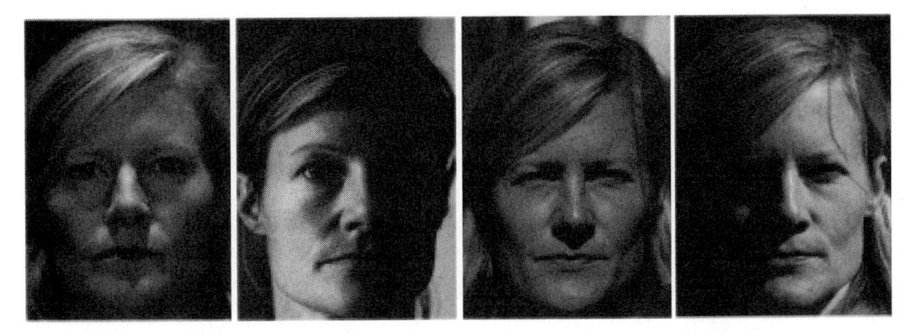

FIGURE 1 The same individual imaged with the same camera and with nearly the same facial expression and pose may appear dramatically different with changes in lighting conditions. The first two images were taken indoors; the third and fourth images were taken outdoors. All four images were taken with a Canon EOS 1D digital camera. Before each picture was taken, the subject was asked to make a neutral facial expression and to look directly into the lens.

"[V]ariations between the images of the same face due to illumination and viewing direction are almost always larger than image variations due to changes in face identity" (Moses et al., 1994). As Figure 1 shows, the same person, with the same facial expression, can appear strikingly different with changes in the direction of the light source and point of view. These variations are exacerbated by additional factors, such as facial expression, perspiration, hair style, cosmetics, and even changes due to aging.

The problem of face recognition can be cast as a standard pattern-classification or machine-learning problem. Imagine we are given a set of images labeled with the person's identity (the gallery set) and a set of images unlabeled from a group of people that includes the individual (the probe set), and we are trying to identify each person in the probe set. This problem can be attacked in three steps. In the first step, the face is located in the image, a process known as face detection, which can be as challenging as face recognition (see Viola and Jones, 2004, and Yang et al., 2000, for more detail). In the second step, a collection of descriptive measurements, known as a feature vector, is extracted from each image. In the third step, a classifier is trained to assign a label with a person's identity to each feature vector. (Note that these classifiers are simply mathematical functions that return an index corresponding to a subject's identity.)

In the last few years, numerous feature-extraction and pattern-classification methods have been proposed for face recognition (Chellappa et al., 1995; Fromherz, 1998; Pentland, 2000; Samil and Iyengar, 1992; Zhao et al., 2003). Geometric, feature-based methods, which have been used for decades, use properties and relations (e.g., distances and angles) between facial features, such as eyes, mouth, nose, and chin, to achieve recognition (Brunelli and Poggio, 1993; Goldstein et al., 1971; Harmon et al., 1978, 1981; Kanade, 1973, 1977; Kaufman and Breeding, 1976; Li and Lu, 1999; Samil and Iyengar, 1992; Wiskott et al., 1997). Despite their economical representation and insensitivity to small variations in illumination and point of view, feature-based methods are quite sensitive to the feature-extraction and measurement process, the reliability of which has been called into question (Cox et al., 1996) In addition, some have argued that face recognition based on inferring identity by the geometric relations among local image features is not always effective (Brunelli and Poggio, 1993).

In the last decade, appearance-based methods have been introduced that use low-dimensional representations of images of objects or faces (e.g., Hallinan, 1994; Hallinan et al., 1999; Moghaddam and Pentland, 1997; Murase and Nayar, 1995; Pentland et al., 1994; Poggio and Sung, 1994; Sirovitch and Kirby, 1987; Turk and Pentland, 1991). Appearance-based methods differ from feature-based techniques in that low-dimensional representations are faithful to the original image in a least-squares sense. Techniques such as SLAM (Murase and Nayar, 1995) and Eigenfaces (Turk and Pentland, 1991) have demonstrated that appearance-based methods are both accurate and easy to use. The feature vector used for classification in these systems is a linear projection of the face image in a lower dimensional linear subspace. In extreme cases, the feature vector is chosen as the entire image, with each element in the feature vector taken from a pixel in the image.

Despite their success, many appearance-based methods have a serious drawback. Recognition of a face under particular lighting conditions, in a particular pose, and with a particular expression is reliable *only if the face has been previously seen under similar circumstances.* In fact, variations in appearance between images of the same person confound appearance-based methods. To demonstrate just how severe this variability can be, an array of images (Figure 2) shows variability in the Cartesian product of pose × lighting for a single individual.

If the gallery set contains a very large number of images of each subject in many different poses, lighting conditions, and with many different facial expressions, even the simplest appearance-based classifier might perform well. However, there are usually only a few gallery images per person from which the classifier must learn to discriminate between individuals.

In an effort to overcome this shortcoming, there has been a recent surge in work on 3-D face recognition. The idea of these systems is to build face-recognition systems that use a handful of images acquired at enrollment time to estimate models of the 3-D shape of each face. The 3-D models can then be used to render images of each face synthetically in novel poses and lighting conditions—effectively expanding the gallery set for each face. Alternatively, 3-D models can be used in an iterative fitting process in which the model for each face is rotated, aligned, and synthetically illuminated to match the probe image. Conversely, the models can be used to warp a probe image of a face back to a canonical frontal point of view and lighting condition. In both of these cases, the identity chosen corresponds to the model with the best fit.

The 3-D models of the face shape can be estimated by a variety of methods. In the simplest methods, the face shape is assumed to be a generic average of a large collection of sample face shapes acquired from laser range scans. Georghiades and colleagues (1999, 2000) estimated the face shape from changes in the shading in multiple enrollment images of the same face under varying lighting conditions. Kukula (2004) estimated the shape using binocular stereopsis on two enrollment images taken from slightly different points of view. Ohlhorst (2005) based the estimate on deformations in the grid pattern of infrared light projected onto the face. In another study, Blanz and Vetter (2003) inferred the 3-D face shape from the shading in a single image using a parametric model of face shape. Often, a "bootstrap" set of prior training data of face shape and reflectance taken from individuals who are not in the gallery or probe sets is used to improve the shape and reflectance estimation process.

The 3-D face recognition techniques described above constitute just a small sampling of the work going on in this area, much of it too new to appear in surveys. To give the reader an idea of the potential of these approaches, the seven images (Figure 3, top row) under variable lighting conditions are used to estimate the face shape and reflectance. The estimate is then used to synthesize images of the face (Figure 4) under the same conditions as those shown in Figure 2. Note that much of the variation in appearance in pose and lighting can be inferred from as few as nine gallery images.

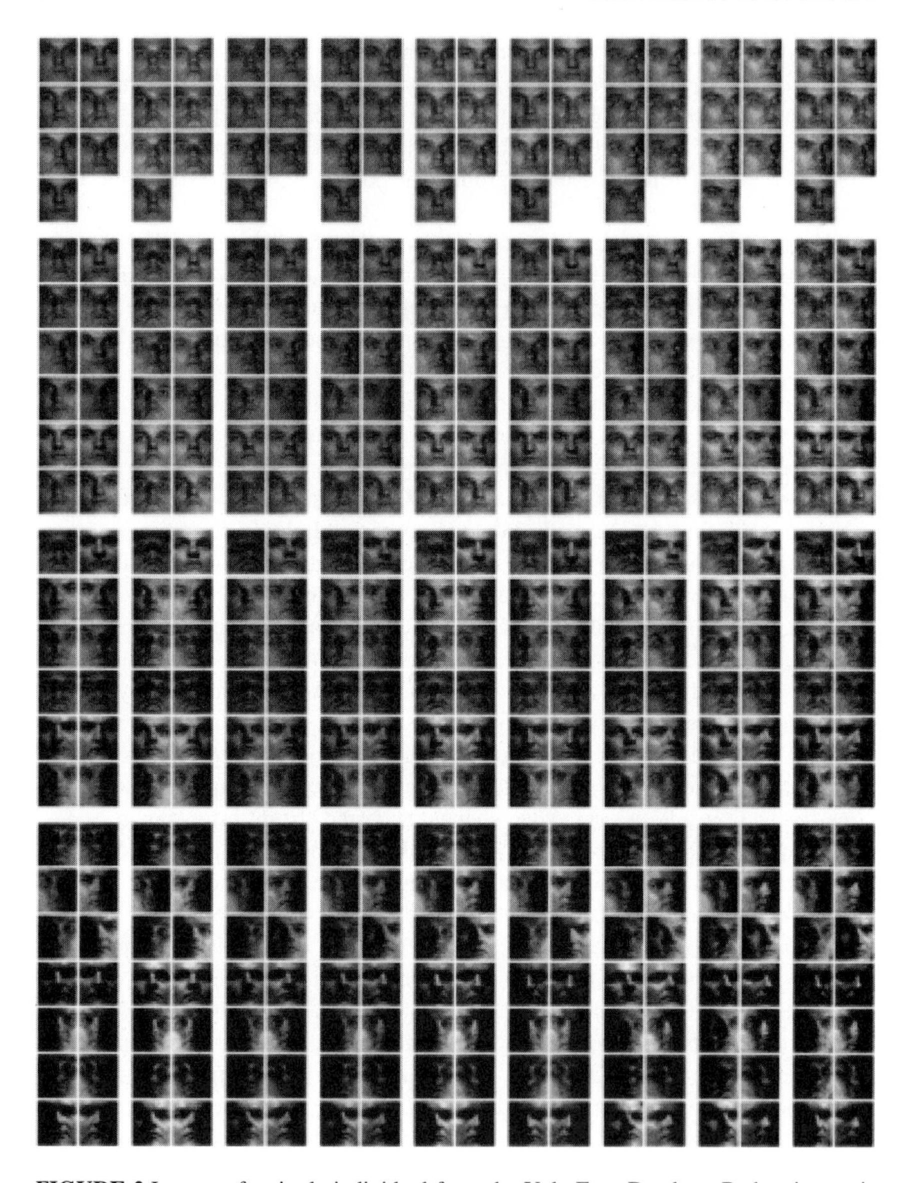

FIGURE 2 Images of a single individual from the Yale Face Database B showing variability due to differences in illumination and pose. The images are divided into four subsets (1 through 4 from top to bottom) according to the angle formed by the light source and the camera axis. Every pair of columns shows the images of a particular pose (1 through 9 from left to right).

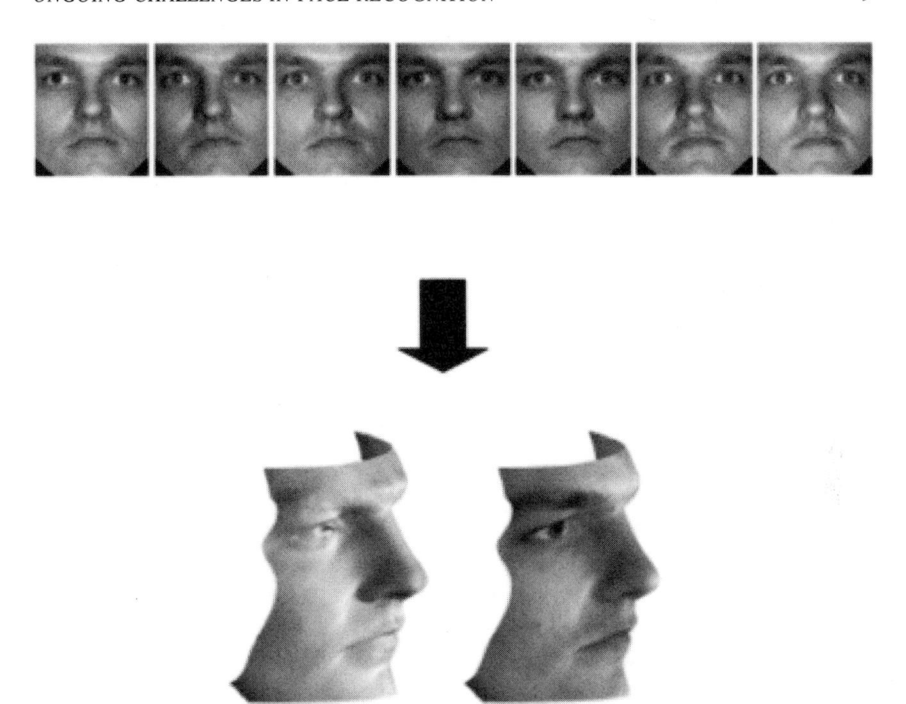

FIGURE 3 A variation of photometric stereopsis was used to compute the shape and reflectance of the face in the bottom row based on the seven gallery images in the top row. Source: Georghiades et al., 1999. Reprinted with permission.

Although recent advances in 3-D face recognition have gone a long way toward addressing the complications causes by changes in pose and lighting, a great deal remains to be done. Natural outdoor lighting makes face recognition difficult, not simply because of the strong shadows cast by a light source such as the sun, but also because subjects tend to distort their faces when illuminated by a strong light; compare again the indoor and outdoor expressions of the subject in Figure 1. Furthermore, very little work has been done to address complications arising from voluntary changes in facial expression, the use of eyewear, and the more subtle effects of aging. The hope, of course, is that many of these effects can be modeled in much the same way as face shape and reflectance and that recognition will continue to improve in the coming decade.

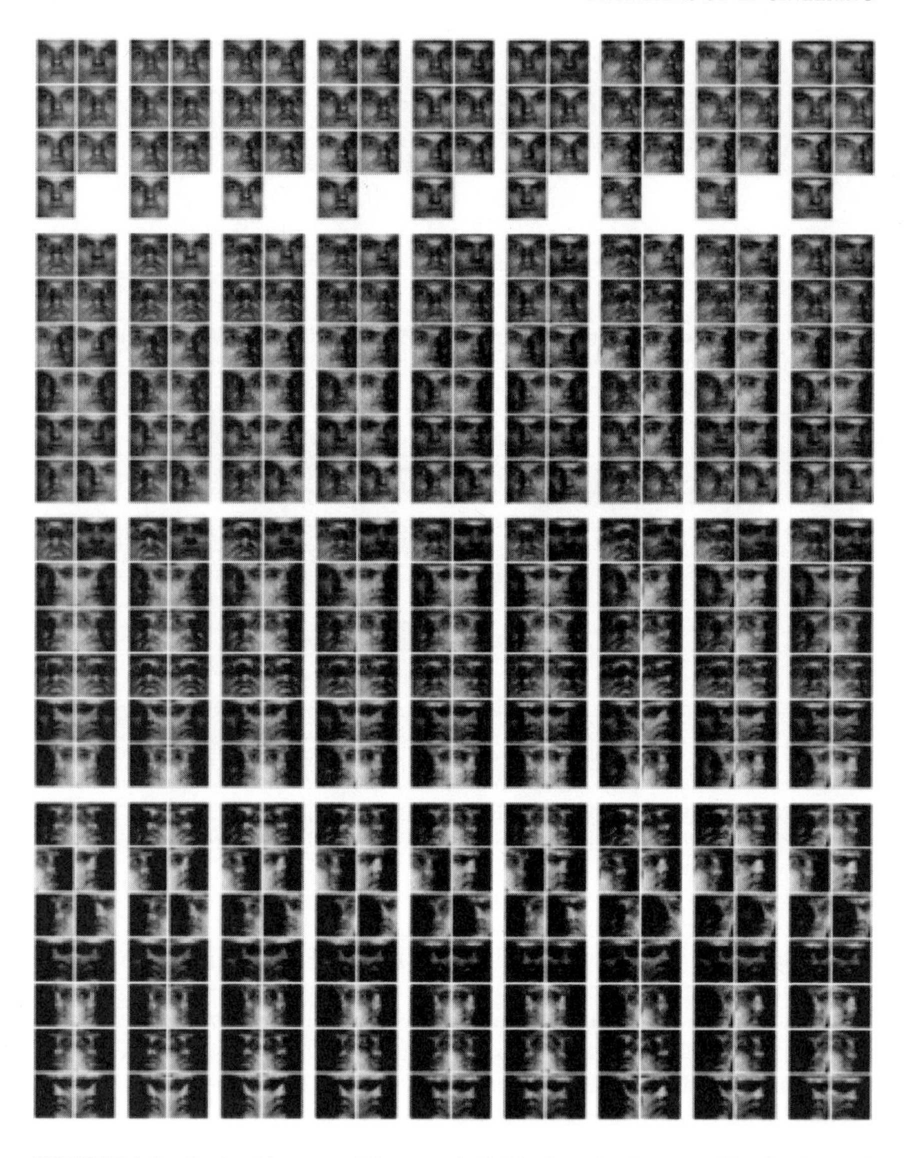

FIGURE 4 Synthesized images of the same individual, under the same illumination and from the same point of view as in Figure 2. Once again, the synthesized images are divided into four subsets (1 through 4 from top to bottom) according to the angle of the light source direction and the camera axis. Every pair of columns shows the images from a particular pose (1 through 9 from left to right). Note that all of the images were generated synthetically from seven gallery images with frontal pose.

REFERENCES

Blanz, V., and T. Vetter. 2003. Face recognition based on fitting a 3D morphable model. IEEE Transactions in Pattern Analysis and Machine Intelligence 25(9): 1063–1074.

Brunelli, R., and T. Poggio. 1993. Face recognition: features vs templates. IEEE Transactions in Pattern Analysis and Machine Intelligence 15(10): 1042–1053.

Chellappa, R., C. Wilson, and S. Sirohey. 1995. Human and machine recognition of faces: a survey. Proceedings of the IEEE 83(5): 705–740.

Cox, I., J. Ghosn, and P. Yianilos. 1996. Feature-Based Face Recognition Using Mixture Distance. Pp. 209–216 in Proceedings of the IEEE Conference on Computer Vision and Pattern Recognition. New York: IEEE.

Fromherz, T. 1998. Face Recognition: A Summary of 1995–1997. ICSI TR-98-027. Berkeley, Calif.: International Computer Science Institute, University of California, Berkeley.

Georghiades, A., P. Belhumeur, and D. Kriegman. 1999. Illumination-Based Image Synthesis: Creating Novel Images of Human Faces under Differing Poses and Lighting. Pp. 47–54 in Proceedings of the IEEE Workshop on Multi-View Modeling and Analysis of Visual Scenes. New York: IEEE.

Georghiades, A., P. Belhumeur, and D. Kriegman. 2000. From Few to Many: Generative Models for Recognition under Variable Poses and Illumination. Pp. 277–284 in Proceedings of the IEEE International Conference on Automatic Face and Gesture Recognition. New York: IEEE.

Georghiades, A., P. Belhumeur, and D. Kriegman. 2001. From Few to Many: Illumination Cone Models for Face Recognition under Variable Lighting and Poses. IEEE Transactions in Pattern Analysis and Machine Intelligence 23(6): 643–660.

Goldstein, A., L. Harmon, and A. Lesk. 1971. Identification of human faces. Proceedings of the IEEE 59(5): 748–760.

Hallinan, P. 1994. A Low-Dimensional Representation of Human Faces for Arbitrary Lighting Conditions. Pp. 995–999 in Proceedings of the IEEE Conference on Computer Vision and Pattern Recognition. New York: IEEE.

Hallinan, P. 1995. A Deformable Model for Face Recognition under Arbitrary Lighting Conditions. Ph.D. thesis, Harvard University. Unpublished.

Hallinan, P., G. Gordon, A. Yuille, and D. Mumford. 1999. Two- and Three-Dimensional Patterns of the Face. Wellesley, Mass.: A.K. Peters.

Harmon, L., S. Kuo, P. Ramig, and U. Raudkivi. 1978. Identification of human face profiles by computer. Pattern Recognition 10(5-6): 301–312.

Harmon, L., M. Kaun, R. Lasch, and P. Ramig. 1981. Machine identification of human faces. Pattern Recognition 13(2): 97–110.

Kanade, T. 1973. Picture Processing by Computer Complex and Recognition of Human Faces. Ph.D. thesis, Kyoto University. Unpublished.

Kanade, T. 1977. Computer Recognition of Human Faces. Stuttgart, Germany: Birkhauser Verlag.

Kaufman, G., and K. Breeding. 1976. The automatic recognition of human faces from profile silhouettes. IEEE Transactions on Systems, Man and Cybernetics 6(February): 113–121.

Kukula, E. 2004. Effects of Light Direction on the Performance of Geometrix FaceVision® 3D Face Recogntion System. Proceedings of the Biometrics Consortium Conference. Gaithersburg, MD: NIST. Available online at: *http://www.biometrics.org/bc2004/Presentations/Conference/2%20 Tuesday%20September%2021/Tue_Ballroom%20A/7%20Purdue%20University%20Session/ Kukula%20-%20BC2004.pdf.*

Li, S., and J. Lu. 1999. Face recognition using nearest feature line. IEEE Transactions on Neural Networks 10(2): 439–443.

Moghaddam, B., and A. Pentland. 1997. Probabilistic visual learning for object representation. IEEE Transactions on Pattern Analysis and Machine Intelligence 19(7): 696–710.

Murase, H., and S. Nayar. 1995. Visual learning and recognition of 3-D objects from appearance. International Journal of Computer Vision 14(1): 5–24.

Ohlhorst, F. 2005. Biometrics Security Solutions Are Here from the Future. Available online at: *http://www.crn.com/showArticle.jhtml?articleID=161601680&flatPage=true* (April 26, 2005).

Pentland, A. 2000. Looking at people: sensing for ubiquitous and wearable computing. IEEE Transactions on Pattern Analysis and Machine Intelligence 22(1): 107–119.

Pentland, A., B. Moghaddam, and T. Starner. 1994. View-Based and Modular Eigenspaces for Face Recognition. Pp. 84–91 in Proceedings of the IEEE Conference on Computer Vision and Pattern Recognition. New York: IEEE.

Poggio, T., and K. Sung. 1994. Example-based learning for view-based human face detection. Pp. II: 843–850 in Proceedings of the ARPA Image Understanding Workshop. Arlington, Va.: DOD.

Samal, A., and P. Iyengar. 1992. Automatic recognition and analysis of human faces and facial expressions: a survey. Pattern Recognition 25(1): 65–77.

Sirovich, L., and M. Kirby. 1987. Low-dimensional procedure for the characterization of human faces. Journal of the Optical Society of America A4(3): 519–524.

Turk, M., and A. Pentland. 1991. Eigenfaces for recognition. Journal of Cognitive Neuroscience 3(1): 71–96.

Viola, P., and M.J. Jones. 2004. Robust real-time face detection. International Journal of. Computer Vision 57(2): 137–154.

Wiskott, L., J. Fellous, N. Kruger, and C. von der Malsburg. 1997. Face recognition by elastic bunch graph matching. IEEE Transactions on Pattern Analysis and Machine Intelligence 19(7): 775–779.

Yang, M., N. Ahuja, and D. Kriegman. 2000. Face Detection Using a Mixture of Linear Subspaces. Pp. 70-76 in Proceedings of the IEEE International Conference on Automatic Face and Gesture Recognition. New York: IEEE.

Zhao, W., R. Chellappa, P.J. Phillips, and A. Rosenfeld. 2003. Face recognition: a literature survey. Association for Computing Machinery Computer Survey 35(4): 399–458.

BIBLIOGRAPHY

Belhumeur, P.N., and D. Kriegman. 1998. What is the set of images of an object under all possible illumination conditions? International Journal of Computer Vision 28(3): 245–260.

Belhumeur, P.N., J.P. Hespanha, and D J. Kriegman. 1997. Eigenfaces vs. Fisherfaces: recognition using class-specific linear projection. IEEE Transactions on Pattern Analysis and Machine Intelligence 19(7): 711–720. Special Issue on Face Recognition.

Belhumeur, P.N., D. Kriegman, and A. Yuille. 1997. The Bas-Relief Ambiguity. Pp. 1040–1046 in Proceedings of the IEEE Conference on Computer Vision and Pattern Recognition. New York: IEEE.

Beymer, D. 1994. Face Recognition under Varying Poses. Pp. 756–761 in Proceedings of the IEEE Conference on Computer Vision and Pattern Recognition. New York: IEEE.

Beymer, D., and T. Poggio. 1995. Face Recognition from One Example View. Pp. 500–507 in Proceedings of the IEEE International Conference on Computer Vision. New York: IEEE.

Blanz, V., and T. Vetter. 1999. A Morphable Model for the Synthesis of 3D Faces. Pp. 187–194 in Proceedings of the 26th Computer Graphics and Interactive Techniques/International Conference on Computer Graphics and Interactive Techniques (SIGGRAPH 1999). New York: ACM Press/Addison-Wesley.

Chen, H., P. Belhumeur, and D. Jacobs. 2000. In Search of Illumination Invariants. Pp. 254–261 in Proceedings of the IEEE Conference on Computer Vision and Pattern Recognition. New York: IEEE.

Chen, Q., H. Wu, and M. Yachida. 1995. Face Detection by Fuzzy Pattern Matching. Pp. 591–596 in Proceedings of the IEEE International Conference on Computer Vision. New York: IEEE.

Cootes, T., G. Edwards, and C. Taylor. 1998. Active Appearance Models. Pp. 484–498 in Proceedings of the European Conference on Computer Vision, vol. 2. Berlin: Springer.

Cootes, T., K. Walker, and C. Taylor. 2000. View-Based Active Appearance Models. Pp. 227–232 in Proceedings of the IEEE International Conference on Automatic Face and Gesture Recognition. New York: IEEE.

Craw, I., D. Tock, and A. Bennet. 1992. Finding Face Features. Pp. 92–96 in Proceedings of the European Conference on Computer Vision. Berlin: Springer.

Edwards, G., T. Cootes, and C. Taylor. 1999. Advances in Active Appearance Models. Pp. 137–142 in Proceedings of the IEEE International Conference on Computer Vision. New York: IEEE.

Epstein, R., P. Hallinan, and A. Yuille. 1995. 5±2 Eigenimages Suffice: An Empirical Investigation of Low-Dimensional Lighting Models. In IEEE Physics Based Modeling Workshop in Computer Vision, Session 4. New York: IEEE.

Frankot, R.T., and R. Chellapa. 1988. A method for enforcing integrability in shape from shading algorithms. IEEE Transactions on Pattern Analysis and Machine Intelligence 10(4): 439–451.

Georghiades, A., D. Kriegman, and P. Belhumeur. 1998. Illumination Cones for Recognition under Variable Lighting: Faces. Pp. 52–59 in Proceedings of the IEEE Conference on Computer Vision and Pattern Recognition. New York: IEEE.

Govindaraju, V. 1996. Locating human faces in photographs. International Journal of Computer Vision 19(2): 129–146.

Hayakawa, H. 1994. Photometric stereo under a light-source with arbitrary motion. Journal of the Optical Society of America 11A(11): 3079–3089.

Horn, B. 1986. Computer Vision. Cambridge, Mass.: MIT Press.

Huang, F., Z. Zhou, H. Zhang, and T. Chen. 2000. Pose Invariant Face Recognition. Pp. 245–250 in Proceedings of the IEEE International Conference on Automatic Face and Gesture Recognition. New York: IEEE.

Jacobs, D. 1997. Linear Fitting with Missing Data: Applications to Structure from Motion and Characterizing Intensity Images. Pp. 206-212 in Procedings of the IEEE Conference on Computer Vision and Pattern Recognition. New York: IEEE.

Juell, P., and R. Marsh. 1996. A hierarchical neural-network for human face detection. Pattern Recognition 29(5): 781–787.

Lambert, J. 1760. Photometria Sive de Mensura et Gradibus Luminus, Colorum et Umbrae. Augsberg, Germany: Eberhard Klett.

Lanitis, A., C. Taylor, and T. Cootes. 1995. A Unified Approach to Coding and Interpreting Face Images. Pp. 368–373 in Proceedings of the IEEE International Conference on Computer Vision. New York: IEEE.

Lanitis, A., C. Taylor, and T. Cootes. 1997. Automatic interpretation and coding of face images using flexible models. IEEE Transactions in Pattern Analysis and Machine Intelligence 19(7): 743–756.

Lee, C., J. Kim, and K. Park. 1996. Automatic human face location in a complex background using motion and color information. Pattern Recognition 29(11): 1877–1889.

Leung, T., M. Burl, and P. Perona. 1995. Finding Faces in Cluttered Scenes Using Labeled Random Graph Matching. Pp. 537–644 in Proceedings of the IEEE International Conference on Computer Vision. New York: IEEE.

Li, Y., S. Gong, and H. Liddell. 2000. Support Vector Regression and Classification Based Multiview Face Detection and Recognition. Pp. 300–305 in Proceedings of the IEEE International Conference on Automatic Face and Gesture Recognition. New York: IEEE.

Moghaddam, B., and A. Pentland. 1995. Probabilistic Visual Learning for Object Detection. Pp. 786–793 in Proceedings of the IEEE International Conference on Computer Vision. New York: IEEE.

Moses, Y., Y. Adini, and S. Ullman. 1994. Face Recognition: The Problem of Compensating for Changes in Illumination Direction. Pp. 286–296 in Proceedings of the European Conference on Computer Vision. Berlin: Springer.

Nayar, S., and H. Murase. 1996. Dimensionality of Illumination Manifolds in Appearance Matching. P. 165 in Proceedings of the International Workshop on Object Representation and Computer Vision. London, UK: Springer-Verlag.

O'Toole, A., T. Vetter, N. Toje, and H. Bulthoff. 1997. Sex classification is better with three-dimensional head structure than with texture. Perception 26: 75–84.

Phillips, P., H. Moon, P. Rauss, and S. Risvi. 1997. The FERET Evaluation Methodology for Face-Recognition Algorithms. Pp. 137–143 in Proceedings of the IEEE Conference on Computer Vision and Pattern Recognition. New York: IEEE.

Phillips, P., H. Wechsler, J. Huang, and P. Rauss. 1998. The FERET database and evaluation procedure for face recognition algorithms. Image and Visual Computing 16(5): 295-306.

Riklin-Raviv, T., and A. Shashua. 1999. The Quotient Image: Class Based Recognition and Synthesis under Varying Illumination Conditions. Pp. II: 566–571 in Proceedings of the IEEE Conference on Computer Vision and Pattern Recognition. New York: IEEE.

Rowley, H., S. Baluja, and T. Kanade. 1998. Neural network-based face detection. IEEE Transactions on Pattern Analysis and Machine Intelligence 20(1): 23–38.

Rowley, H., S. Baluja, and T. Kanade. 1998. Rotation Invariant Neural Network-Based Face Detection. Pp. 38–44 in Proceedings of the IEEE Conference on Computer Vision and Pattern Recognition. New York: IEEE.

Samil, A., and P. Iyengar. 1995. Human face detection using silhouettes. Pattern Recognition and Artificial Intelligence 9: 845–867.

Shashua, A. 1992. Geometry and Photometry in 3D Visual Recognition. Ph.D. thesis, Massachusetts Institute of Technology. Unpublished.

Shashua, A. 1997. On photometric issues in 3D Visual Recognition from a Single 2D Image. International Journal of Computer Vision 21(1–2): 99–122.

Shum, H., K. Ikeuchi, and R. Reddy. 1995. Principal component analysis with missing data and its application to polyhedral object modeling. IEEE Transactions in Pattern Analysis and Machine Intelligence 17(9): 854–867.

Silver, W. 1980. Determining Shape and Reflectance Using Multiple Images. Ph.D. thesis, Massachusetts Institute of Technology. Unpublished.

Tomasi, C., and T. Kanade. 1992. Shape and motion from image streams under orthography: a factorization method. International Journal of Computer Vision 9(2): 137–154.

Vetter, T. 1998. Synthesis of novel views from a single face image. International Journal of Computer Vision 28(2): 103–116.

Vetter, T., and T. Poggio. 1997. Linear object classes and image synthesis from a single example image. IEEE Transactions on Pattern Analysis and Machine Intelligence 19(7): 733–742.

Woodham, R. 1981. Analysing images of curved surfaces. Artificial Intelligence 17: 117–140.

Yang, M., D.J. Kriegman, and N. Ahuja. 2002. Detecting faces in images: a survey. IEEE Transactions on Pattern Analysis and Machine Intelligence 24(1): 34–58.

Yow, K., and R. Cipolla. 1997. Feature-based human face detection. Image Visual Computing 15(9): 713–735.

Yu, Y., and J. Malik. 1998. Recovering photometric properties of architectural scenes from photographs. Computer Graphics (SIGGRAPH 1998 Conference Proceedings) 32: 207–217.

Yuille, A., and D. Snow. 1997. Shape and Albedo from Multiple Images Using Integrability. Pp. 158–164 in Proceedings of the IEEE Conference on Computer Vision and Pattern Recognition. New York: IEEE.

Zhao, W., and R. Chellapa. 2000. SFS Based View Synthesis for Robust Face Recognition. Pp. 285–292 in Proceedings of the IEEE International Conference on Automatic Face and Gesture Recognition. New York: IEEE.

Zhao, W., R. Chellappa, and P. Phillips. 1999. Subspace Linear Discriminant Analysis for Face Recognition. CAR-TR-914. College Park, Maryland: Center for Automation Research, University of Maryland.

Designing Biometric Evaluations and Challenge Problems for Face-Recognition Systems

P. Jonathon Phillips
National Institute of Standards and Technology
Gaithersburg, Maryland

Automatic face recognition is the only area of computer vision and pattern recognition with more than a decade of history of challenge problems and independent evaluations. These challenge problems have provided the face-recognition community with a large corpus of data for algorithm and technology development. Periodic evaluations have been conducted to measure progress in the performance of these algorithms and technologies. As face-recognition technology has matured, so has the sophistication of challenge problems and evaluations.

Prior to the first face-recognition evaluations, researchers reported performance on small propriety databases, usually of fewer than 100 people, for partially automatic algorithms. Partially automatic algorithms require that the location of the eyes (a "ground truth") be provided; fully automatic algorithms process facial images without manual intervention or ground truths. From self-reported results on propriety data sets, it was not possible to make objective comparisons of different approaches or to access the best techniques.

The tradition of challenge problems and evaluations in face recognition began with the FERET Program, which ran from 1993 to 1997. Under this program, a large data set was collected and then partitioned into two sets—a development set and a sequestered set. The development set of images was made available to researchers for algorithm development. The sequestered set, as the name implies, was sequestered and used to make independent evaluations of algorithm performance on images they had not seen before. Because all algorithms are tested on exactly the same data, it is possible to make direct comparisons of performance with different algorithms.

The initial goal of the FERET Program was to determine if automatic face recognition was possible. This question was answered in the affirmative by the August 1994 evaluation (Phillips et al., 1998). Two more FERET evaluations were conducted, in March 1995 and September 1996, to measure progress under the FERET Program (Phillips et al., 2000). The last FERET evaluation, in September 1996, measured performance on a data set of 1,196 people and 3,323 images.

The FERET evaluations showed significant advances in the development of face-recognition technology. The FERET evaluations addressed the following basic problems: (1) the effect on performance of gallery size and (2) the effect on performance of temporal variations. At the conclusion of FERET, state-of-the-art algorithms were fully automatic, could process 3,815 images of 1,196 people, and could recognize the faces of people from pairs of facial images taken 18 months apart.

In biometrics, including face recognition, performance is reported for three types of tasks—verification, identification, and watch list tasks. A verification task asks, "Am I who I say I am?" An identification task asks, "Who am I?" And a watch list task asks, "Am I someone you are looking for?"

In a verification task, a person presents his or her biometric and an identity claim to a face-recognition system. The system then compares the presented biometric with a stored biometric of the claimed identity. Based on the results of the comparison between the new and stored biometric, the system either accepts or rejects the claim. There are two types of system users—legitimate users and persons who attempt to impersonate legitimate users. Verification performance is characterized by two performance statistics that show the success rate for the two types of users. The *verification rate* is the rate at which legitimate users are granted access. The *false accept rate* (FAR) is the rate at which imposters are granted access. An ideal system has a verification rate of 100 percent and a FAR of 0 percent.

Unfortunately, no ideal system exists. In real-world systems, there is always a trade-off between the verification rate and the FAR. Therefore, it is critical that FARs and verification rates be considered together in determining the performance capabilities of a face-recognition system. It is easy to build a system that always grants access to a subject. This system will have a 100 percent verification rate because access will always be granted in response to a legitimate user's request. Conversely, this system will also have a 100 percent FAR because it also grants access to imposters. The best system is one that balances the verification rate with the FAR in a manner consistent with operational needs.

FRVT 2002

The successor to the FERET series of evaluations is the Face Recognition Vendor Test (FRVT) series of evaluations. The primary objective of FRVT 2002, a large-scale evaluation of automatic face-recognition technology, was to provide performance measures against real-world requirements. FRVT 2002 measured performance of the core capabilities of face-recognition technology and provided an assessment of the potential of face-recognition technology for meeting the requirements for operational applications (Phillips et al., 2003).

FRVT 2002 was independently administered to ten participants that were evaluated under the direct supervision of the FRVT 2002 organizers at a U.S. government facility in Dahlgren, Virginia, in July and August 2002. Participants were tested using data that they had not seen previously. The heart of the FRVT 2002 was the high computational intensity test (HCInt), which consisted of 121,589 operational images of 37,437 people from the U.S. Department of State Mexican Non-Immigrant Visa Archive. From these data, real-world performance figures on a very large data set were computed for verification, identification, and watch list tasks.

The most likely application of face-recognition technology would use images taken indoors. FRVT 2002 results show that normal changes in indoor lighting do not significantly affect the performance of the top systems. In FRVT 2000, the results obtained using two indoor data sets with different lighting were approximately the same. In both experiments, the best performer had a 90 percent verification rate and a FAR of 1 percent.

For the best face-recognition systems, the recognition rate for faces captured *outdoors,* at a FAR of 1 percent, was 50 percent. Thus, face recognition from outdoor imagery remains a research challenge area. The FRVT 2002 database also consisted of images of the same person taken on different days. The performance results for indoor images showed that the capabilities of face-recognition systems had improved over similar experiments conducted two years earlier in FRVT 2000. The results of FRVT 2002 indicated that there had been a 50 percent reduction in error rates.

A very important question for real-world applications is the rate of decrease in performance as the time interval increases between the acquisition of the database images and new images presented to a system. For the top systems, performance degraded at approximately 5 percentage points per year.

One open question is still how the size of the database and the size of the watch list affect performance. Because of the large number of people and images in the FRVT 2002 data set, we were able to report the first large-scale results on this question. For the best system, the top-rank identification rate was 85 percent on a database of 800 people, 83 percent on a database of 1,600, and 73 percent on a database of 37,437. For every doubling of the size of the database, performance decreased by 2 to 3 overall percentage points. In mathematical terms, identification performance decreased linearly with respect to the logarithm of the database size.

A similar effect was observed for the watch list task. As the watch list size increased, performance decreased. For the best system, the identification and detection rate was 77 percent at a FAR of 1 percent for a watch list of 25 people. For a watch list of 300 people, the identification and detection rate was 69 percent at a FAR of 1 percent. In general, systems performed better with a watch list of 25 to 50 people than with a longer watch list.

Previous evaluations had reported face-recognition performance as a function of imaging properties. For example, they compared the differences in performance for indoor and outdoor images, or frontal and non-frontal images. FRVT 2002, for the first time, considered the effects of demographics on performance. This revealed two major effects. First, recognition rates for male images were higher than for female images. For the top systems, identification rates for male images were 6 to 9 percentage points higher than for female images. For the best system, identification rates for male images was 78 percent and for females 69 percent. Second, recognition rates for images of older people were higher than for images of younger people. For 18 to 22 year olds, the average identification rate for the top systems was 62 percent, and for 38 to 42 year olds it was 74 percent. For every ten-year increase in age, average performance improved approximately 5 percent through age 63. All identification rates were computed from a database of 37,437 individuals.

Since FRVT 2000, new techniques and approaches to face recognition have emerged. Two of these new techniques were evaluated in FRVT 2002. The first was a three-dimensional morphable model technique developed by Blanz and Vetter (1999) to improve recognition of non-frontal images. We found that Blanz and Vetter's technique significantly improved recognition performance. The second technique was recognition from video sequences. Using FRVT 2002 data sets, we found that recognition levels using video sequences was the same as with still images.

In summary, several key lessons were learned from FRVT 2002:

- Given reasonable, controlled indoor lighting, the current state of the art in face recognition is 90 percent verification at a 1 percent FAR.
- The use of morphable models can significantly improve non-frontal face recognition.
- Watch list performance decreases as a function of size—performance for smaller watch lists is better than performance for larger watch lists.
- Demographic characteristics, such as age and sex, can significantly affect performance. Therefore, accommodations should be made for demographic information.

FACE RECOGNITION GRAND CHALLENGE

In the last few years, researchers have been developing new techniques fueled by advances in computer-vision techniques, computer design, sensor design, and the growing interest in fielding face-recognition systems. Proposed new techniques include recognition from three-dimensional (3-D) scans, recognition from high-resolution still images, recognition from multiple still images, multi-modal face recognition, multi-algorithms, and preprocessing algorithms to correct for variations in illumination and pose. The hope is that these advances will reduce the error rate in face-recognition systems by an order of magnitude over FRVT 2002 (Phillips et al., 2003).

The Face Recognition Grand Challenge (FRGC), a technology-development project at the National Institute of Standards and Technology, is designed to achieve this performance goal by pursuing the development of algorithms for all of the proposed methods. To facilitate the development of new algorithms, a data corpus of 50,000 recordings, divided into training and validation partitions, was provided to researchers. The images consisted of 3-D scans and high-resolution still images taken under controlled and uncontrolled conditions. The 3-D scans consist of both shape and texture channels.

A primary goal of the upcoming FRVT 2006 is to determine if the goals of FRGC have been met.[1] The starting point for measuring improvements in performance is the HCInt of FRVT 2002, which used images taken indoors under controlled lighting conditions. The performance point selected as the reference is a verification rate of 80 percent (error rate of 20 percent) at an FAR of 0.1 percent (the performance level of the top three FRVT 2002 participants). An order of magnitude increase in performance would be a verification rate of 98 percent (2 percent error rate) at the same fixed FAR of 0.1 percent.

SUMMARY OF GRAND CHALLENGE PERFORMANCE

Participants in FRGC submitted raw similarity scores to FRGC organizers on January 14, 2005 (for a detailed description of the FRGC challenge problem, data, and experiments, see Phillips et al., 2005). The experiments in FRGC ver2.0 are designed to advance face recognition in general, with an emphasis on 3-D and high-resolution still imagery. Ver2.0 consists of six experiments.

Experiments

Experiment 1 measures performance on the classic face-recognition problem—recognition of a faces from frontal images taken under controlled illumi-

[1]FRVT 2006, which is scheduled to begin on January 30, 2006, is open to academia, research institutions, and companies.

nation. To encourage the development of high-resolution recognition, all controlled still images in this experiment are high resolution. In biometric evaluations, the set of images known to a system is called the *target set*, and the set of unknown images presented to a system is called the *query set*. In Experiment 1, the biometric samples in the target and query sets consist of a single, controlled still image.

Experiment 2 is designed to examine the effect on performance of multiple still images. In this experiment, each biometric sample consists of four controlled images of a person taken in a subject session.

Experiments 3, 5, and 6 look at different potential implementations of 3-D face recognition. Experiment 3 measures performance when both the target and query images are 3-D. There are three versions of Experiment 3. The main version compares both the shape and texture channels of 3-D images. Experiment 3t compares just the texture channels, and Experiment 3s compares just the shape channels. In all versions of Experiment 3, the target and query sets consist of 3-D facial images. One potential scenario for 3-D face recognition is that the target images are 3-D and the query images are two dimensional (2-D) images.

Experiment 4 is designed to measure progress on recognition from uncontrolled frontal still images. In this experiment, the target set consists of single controlled still images, and the query set consists of a single uncontrolled still image.

Experiment 5 looks at the same scenario as Experiment 3 using controlled query images. Experiment 6 looks at the same scenario with uncontrolled query images. In both Experiments 5 and 6, the target set consists of 3-D images.

Table 1 and Figure 1 summarize our results. Table 1 shows the number of similarity matrices analyzed for each experiment. The bar graph in Figure 1 summarizes performance for each experiment by the verification rate at a FAR of 0.001, the vertical axis. Three statistics are reported for each experiment: the performance of the baseline algorithm (left bar); the best performance among the submitted similarity matrices for an experiment (right bar); and the median performance over submitted results for each experiment (center bar). For Experiments 5 and 6, no baseline algorithm was provided and only one result was submitted, which is reported.

The maximum score for Experiment 1 was 99 percent, and the median was 91 percent. The comparable scores for Experiment 2 are 99.9 percent and 99.9 percent. Because FRGC is a challenge problem and the results are based on raw

TABLE 1 Number of Results Submitted for Each Experiment

	Experiment							
	1	2	3	3t	3s	4	5	6
Number of results	17	11	10	4	5	12	1	1

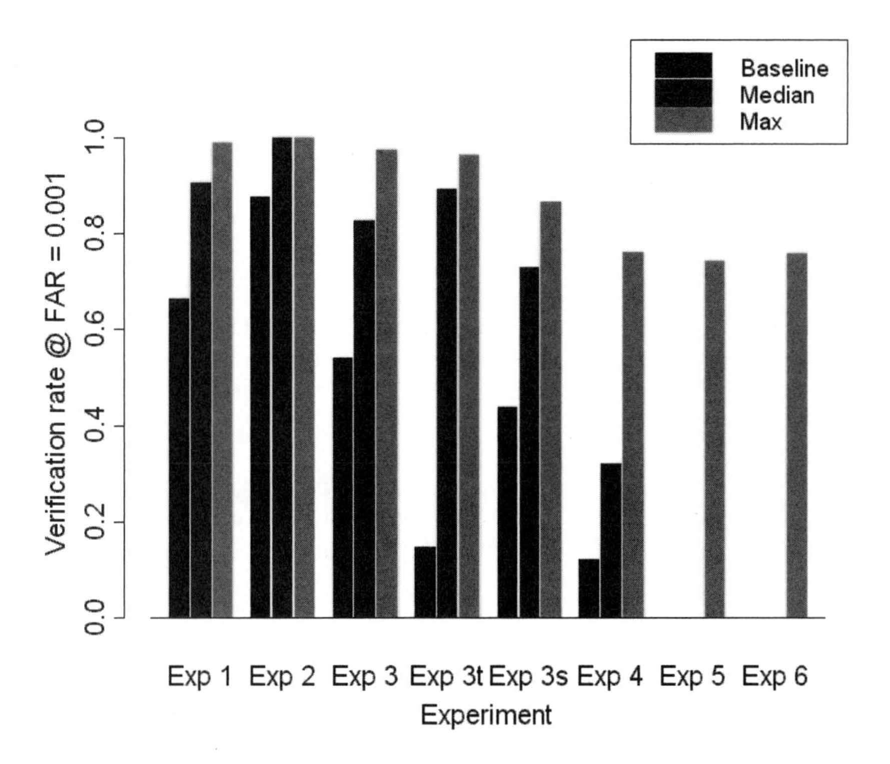

FIGURE 1 Summary of performance results for FRGC Experiments 1, 2, 3, 3t, 3s, 4, 5, and 6.

similarity scores submitted by participating groups, the results are not conclusive that the performance goals of FRGC have been met. However, they do provide evidence that the goals are likely to be met. The difference in performance between the results for Experiments 1 and 2, especially for median scores, indicate that having multiple still images of a person can potentially improve performance.

FRGC is the first time a large set of 3-D facial imagery was made available. The maximum score of 97 percent for Experiment 3 shows the potential of using 3-D facial imagery for face recognition. The results for Experiment 3 were obtained only three months after the first release of a large 3-D data set. By comparison, the results for still images are based on more than a decade of intensive research after the release of the first large still-image data sets.

CONCLUSION

The FERET, FRVT, and FRGC projects and evaluations have been instrumental in advancing automatic face-recognition technology. Prior to FERET, it was not possible to compare competing methods or to make direct comparisons of the effectiveness of different algorithms. FRVT 2002, which supplied reported performance rates on a large data set of operational images, served as a baseline for progress under FRGC. FRGC is facilitating the development of the next generation of face recognition. Progress under FRGC will be measured by FRVT 2006.

REFERENCES

Blanz, V., and T. Vetter. 1999. A Morphable Model for the Synthesis of 3D Faces. Pp. 187–194 in SIGGRAPH '99, Proceedings of the Annual Conference on Computer Graphics, August 8–13, 1999, Los Angeles, California. New York: Association for Computing Machinery.

Phillips, P.J., H. Wechsler, J. Huang, and P. Rauss. 1998. The FERET database and evaluation procedure for face-recognition algorithms. Image and Vision Computing 16(5): 295–306.

Phillips, P.J., H. Moon, S. Rizvi, and P. Rauss. 2000. The FERET evaluation methodology for face-recognition algorithms. IEEE Transactions on Pattern Analysis and Machine Intelligence 22(10): 1090–1104.

Phillips, P.J., P.J. Grother, R.J. Michaels, D.M. Blackburn, E. Tabassi, and J.M. Bone. 2003. Face Recognition Vendor Test 2002: Evaluation Report. Technical Report NISTIR 6965. Gaithersburg, Md.: National Institute of Standards and Technology. Available online at: *http://www.frvt.org*.

Phillips, P.J., P.J. Flynn, T. Scruggs, K.W. Bowyer, J. Chang, K. Hoffman, J. Marques, J. Min, and W. Worek. 2005. Overview of the Face Recognition Grand Challenge. Pp. 947–954 in IEEE Computer Society Conference on Computer Vision and Pattern Recognition. New York: IEEE.

Large-Scale Activity-Recognition Systems

MATTHAI PHILIPOSE
Intel Research Laboratory
Seattle, Washington

Building computing systems that can observe, understand, and act on day-to-day physical human activity has long been a goal of computing research. Such systems could have profound conceptual and practical implications. Because the ability to reason and act based on activity is a central aspect of human intelligence, from a conceptual point of view, such a system could improve computational models of intelligence. More tangibly, machines that can reason about human activity could be useful in aspects of life that are currently considered outside the domain of machines.

Monitoring human activity is a basic aspect of reasoning about activity. In fact, monitoring is something we all do—parents monitor children, adults monitor elderly parents, managers monitor teams, nurses monitor patients, and trainers monitor trainees; people following medication regimens, diets, recipes, or directions monitor themselves.

Besides being ubiquitous, however, monitoring can also be tedious and expensive. In some situations, such as caregiver-caretaker and manager-worker relationships, only dedicated, trained human monitors can make detailed observations of behavior. However, such extensive observation causes fatigue in observers and resentment in those being observed. The constant involvement of humans also makes monitoring expensive.

Tasks that are ubiquitous, tedious, and expensive are usually perfect candidates for automation. Machines do not mind doing tedious work, and expensive problems motivate corporations to build machines. In fact, given the demographics of our society, systems that notify family members automatically when elderly relatives trigger simple alarms, such as falling, not turning off the stove, or not turning off hot water, are now commercially available. However, compared to a live-in family member who can monitor an elder's competence in thousands of day-to-day activities, these systems barely scratch the surface. In this paper, I describe a concrete application for a monitoring system with broad activity-recognition capabilities, identify a crucial missing ingredient in existing activity "recognizers," and describe how a new class of sensors, combined with emerging work in statistical reasoning, promises to advance the state of the art by providing this ingredient.

THE CAREGIVER'S ASSISTANT

Caring for the elderly, either as a professional caregiver or as a family member, is a common burden in most societies. Gerontologists have developed a detailed list of activities, called the activities of daily living (ADLs), and metrics for scoring performance of crucial day-to-day tasks, such as cooking, dressing, toileting, and socializing, which are central to a person's well-being. An elder's ADL score is accepted as an indicator of his or her cognitive health.

Professional caregivers in the United States are often required to fill in ADL forms each time they visit their patients. Unfortunately, although the data they collect are used as a basis for making resourcing decisions, such as Medicaid payments, the data are often inaccurate because (1) they are often based on interviews with elders who may have strong motives for misrepresenting the facts and (2) because the data-collection window is narrow relative to the period being evaluated. Given increasing constraints on caregivers' time, purely manual data collection seems unsustainable in the long run.

The Caregiver's Assistant system is intended to fill out large parts of the ADL form automatically based on data collected from the elder's home on a 24/7 basis. The system would not just improve the quality of data collected, but (because it provides constant monitoring) might also be able to provide proactive intervention and other assistance. Figure 1 shows a prototype form of the Caregiver's Assistant. Actual forms include activities in 23 categories, such as "housework" and "hygiene," which instantiate to tens of thousands of activities, such as "cleaning a bathtub" and "brushing teeth."

Thus, an activity-recognition system that could track thousands of activities in non-laboratory conditions would remove a substantial burden from human monitors. Professional caregivers could, at any time, be provided with a version of this form with potentially troublesome areas highlighted. If a nurse were given this form before a visit, for instance, he or she could make better preparations for the visit and could focus on the most important issues during the visit. A study of roughly one hundred professional caregivers around the country has shown that such a system would be useful, at least for caregivers.

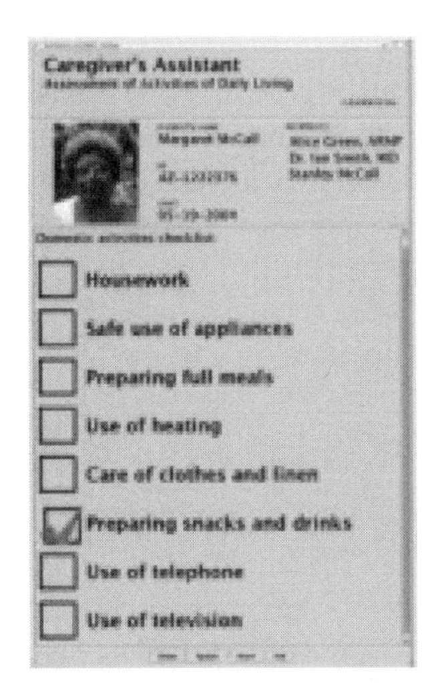
Caregiver's Assistant

Housework
Safe use of appliances
Preparing full meals
Use of heating
Care of clothes and linen
Preparing snacks and drinks
Use of telephone
Use of television

FIGURE 1 Prototype of electronic Activities of Daily Living Form with check mark added by the electronic Caregiver's Assistant.

DISCRIMINATING AMONG ACTIVITIES

The process of recognizing mundane physical activities can be understood as mapping from raw data gathered by sensors to a label denoting an activity. Figure 2 shows how traditional mapping systems are structured. *Feature selection modules* typically work on high-dimensional, high-frequency data coming directly from sensors (such as cameras, microphones, and accelerometers) to

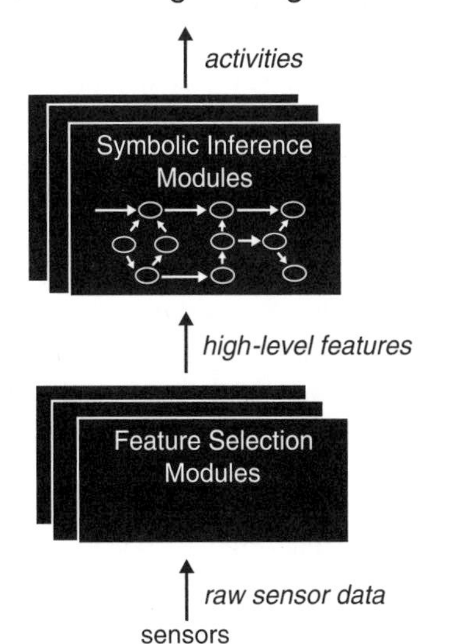

FIGURE 2 A typical activity-recognition system.

identify relatively small numbers of semantically higher level features, such as objects in images, phonemes in audio streams, and motions in accelerometer data. *Symbolic inference modules* reason about the relationship between these features and activities in a variety of ways. The reasoning may include identifying ongoing activities, detecting anomalies in the execution of activities, and performing actions to help achieve the goal of the activities.

Both feature selection and inference techniques have been investigated extensively, and depending on the feature, researchers can draw on large bodies of work. In the computer vision community alone, extensive work has been done on objects, faces, automobiles, gestures, and edges and motion flows, each of which has a dedicated sub-community of researchers. Thus, once features for an activity-recognition system have been selected, a very large number of model representations and inference techniques are available. These techniques differ in several ways, such as whether they support statistical, higher order, or temporal reasoning; the degree to which they learn and the amount of human intervention they require to learn; and the efficiency with which they process various kinds of features, especially higher dimensional features. In Figure 2, the variety of feature selections and inference algorithms is indicated by stacks of boxes.

Despite the profusion of options, no activity inferencing system capable of recognizing large numbers of day-to-day activities in natural environments has been developed. A key underlying problem is that no existing combination of sensors and feature selector has been shown to detect robustly the features necessary to distinguish between thousands of activities. For instance, objects used during activities have long been thought to be crucial discriminators. However, existing object-recognition and tracking systems tend not to work very well when applied to a large variety of objects in unstructured environments (Sanders et al., 2002). Activity-recognition systems based on tracking objects, therefore, tend to be customized for particular environments and objects, which limits their utility as general-purpose, day-to-day activity recognizers. Given that producing each customized detector is a research task, the goal of general-purpose recognition has, not surprisingly, not been reached.

A new class of small, wireless sensors seems likely to provide a practical means of detecting objects used in many day-to-day activities (Philipose et al., 2004; Tapia et al., 2004). Given a stream of objects, recent work has shown that even simple symbolic inference techniques are sufficient for tracking the progress of these activities.

DETECTING OBJECT USE WITH RADIO FREQUENCY IDENTIFICATION TAG SENSORS

A passive radio frequency identification (RFID) tag (Figure 3a) is a postage-stamp-sized, wireless, battery-free transponder that, when interrogated (via radio) by an ambient reader, returns a unique identifier (Finkenzeller, 2003). Each

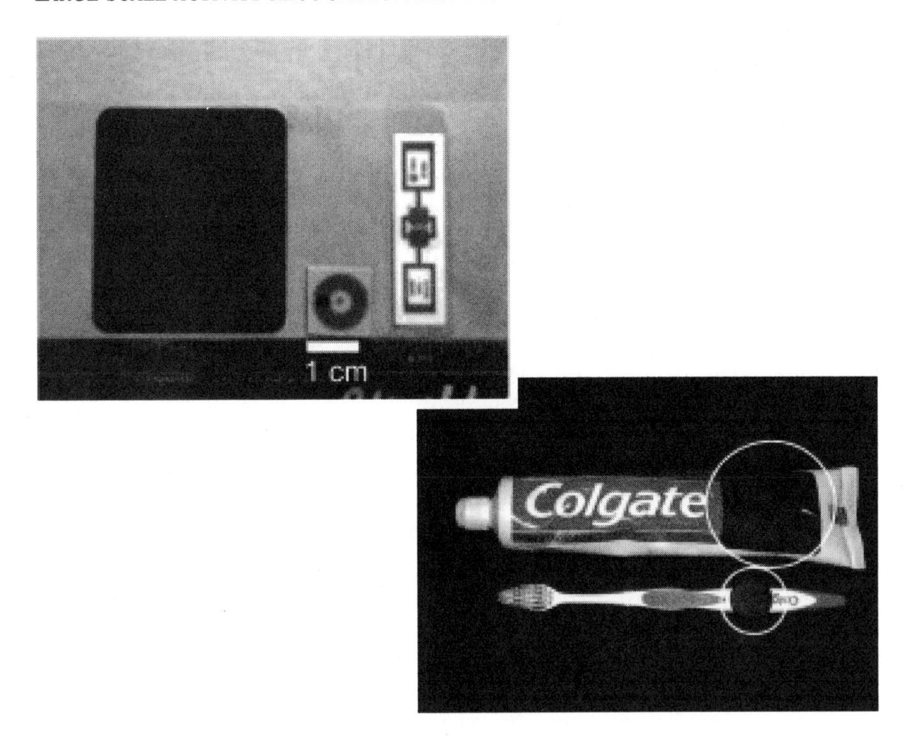

FIGURE 3 a. RFID tags. b. Tagged toothbrush and toothpaste.

tag consists of an antenna, some protocol logic, and optional nonvolatile memory. RFID tags use the energy of the interrogating signal to return a 64-bit to 128-bit identifier unique to each tag, and when applicable, data stored in on-tag memory. Short-range tags, which are inductively coupled, have a range of 2 to 30 cm; long-range backscatter-based tags have a range of 1 to 10 m. Tags are available off the shelf for less than 50 cents each. Short-range readers cost a few hundred dollars; long-range readers cost a few thousand dollars. If current trends continue, there will be a steep drop in the price of both tags and readers in the next few years.

If an RFID tag is attached to an object (Figure 3b) and the tag is detected in the vicinity of a reader, we can infer that the attached object is also present. Given their object-tracking abilities, RFID-based systems are currently being seriously considered for commercial applications, such as supply-chain management and asset tracking. Existing uses include livestock tracking, theft protection in the retail sector, and facilities management. The promise of a viable RFID system for tracking the presence of large numbers of objects suggests that it might be the basis of a system for tracking objects used by people whose activities we wish to monitor. Because a sensor can be attached to each object, we have, in principle at least, an "ultra-dense" deployment of sensors that could allow each tagged object to "report" when it is in use.

However, neither short-range nor long-range RFID systems, as conventionally designed, are quite up to the task of detecting object use in a way that would be useful for tracking activity. Short-range RFID readers are typically bulky hand-held units (similar to bar-code readers) that must be intentionally "swiped" on tags. Clearly, it is not practical to expect a person whose activities are being tracked (whether an elder or a medical student) to carry a scanner and swipe tagged objects in the middle of day-to-day tasks.

Long-range tags, however, do not require the explicit cooperation of those being monitored. Readers in the corner of a room can detect tags anywhere in that room. Unfortunately, because a conventional RFID tag simply reports the *presence of* tagged objects in the reader's field, and not their *use*, long-range tags cannot tell us when objects are being used either. Long-range tags simply list all tagged objects in the room they are monitoring.

However, each of these modalities can be re-engineered to detect object use unobtrusively. Figure 4 shows how the short-range RFID reader can be adapted to become an unobtrusive sensor of object use (Figure 4a). Essentially, the RFID reader is a radio with built-in processor, nonvolatile memory, and a power supply integrated into a single bracelet called the iBracelet (Fishkin et al., forthcoming). The antenna of the RFID reader is built into the rim of the bracelet. When turned on, the bracelet scans for tags at 1 Hz at a range of 20–30 cm. Any object, such as the water pitcher in Figure 4b, that has a tag within 10 to 15 cm of its grasping surface, can therefore be identified as having been touched. The data can either be stored on board (for later offloading through a data port) or imme-

FIGURE 4 a. Close-up of an iBracelet with a quarter for comparison. b. An iBracelet in use.

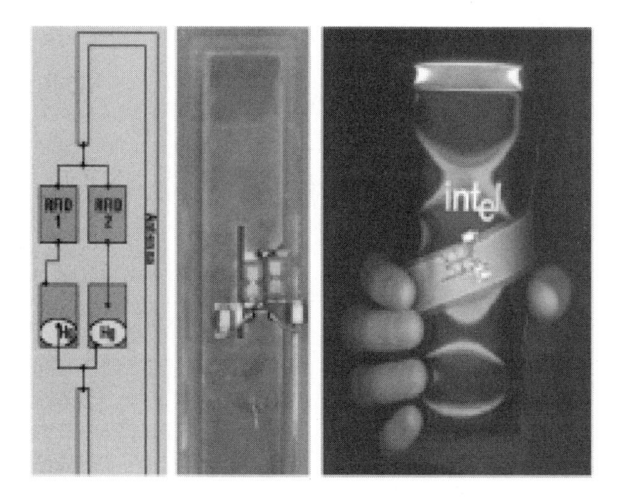

FIGURE 5 WISPs: a. Schematic drawing. b. A single α-WISP. c. A WISP on a coffee mug.

diately radioed off board. The bracelet can currently read for 30 hours between charges when storing data locally, and roughly 10 hours when transmitting data.

Careful placement of tags on objects can reduce false negative rates (i.e., tags being missed). However, given the range of the bracelet, "accidental" swipes of objects are unavoidable. Therefore, the statistical framework that processes the data must be able to cope with these false "hits." Early studies indicate that an iBracelet equipped with inexpensive inductively coupled tags are a practical means of detecting object touch, and therefore object use.

Some people may consider wearing a bracelet an unacceptable requirement, however. In these cases, wireless identification and sensing platforms (WISPs) may be a useful way of detecting object use (Philipose et al., 2005). WISPs, essentially long-range RFID tags with integrated sensors, use incident energy from distant readers not only to return a unique identifier, but also to power the onboard sensor and communicate the current value of the sensor to the reader. For activity-inferencing applications, so-called α-WISPs, which have integrated accelerometers and are about the size of a large Band-Aid™, are attached to objects being tracked (Figure 5). When a tagged object is used, more often than not the accelerometer is triggered and the ambient reader notified.

A single room, which may contain hundreds of tagged objects (most of them inactive at any given time), can be monitored by a single RFID reader. A complication with WISPs is that the explicit correspondence between the person using the object and the object being used is lost. Thus, higher-level inference software may be necessary to track the correspondence implicitly.

INFERENCE SYSTEMS

Given the sequence of objects detected by RFID-based sensors, the job of the inference system is to infer the type of activity. The inference system relies on a model that translates from observations (in this case, the objects seen) to the activity label. Recent work has shown that even very simple statistical models of activities are sufficient to distinguish between dozens of activities performed in a real home (Philipose et al., 2004).

Figure 6 shows a model for making tea, in which each activity is represented as a linear sequence of steps. Each step has a specified average duration, a set of objects likely to be seen in that step, and the probability that one of these objects will be seen in an observation window. In the figure, the first step (corresponding to boiling tea) takes five minutes on average; in each one-second window, there is a 40, 20, and 30 percent chance respectively of a kettle, stove, or faucet being used. Experiments in a real home with 14 subjects, each performing a randomly selected subset of 66 different activities selected from ADL forms, and using activity models constructed by hand to classify the resulting data automatically, have yielded higher than 70 percent (and often close to 90 percent) accuracy in activity detection.

Although the models are simple, it is still impractical to model tens of thousands of activities by hand. However, because the features to be recognized are English words that represent objects and the label to be mapped to is an English phrase (such as "making tea"), the process of building a model is essentially translating probabilistically from English phrases to words. Recent work based on this observation has successfully, completely automatically, extracted translations using word co-occurrence statistics from text corpora, such as the Web (Wyatt et al., 2005). If one million Web pages mention "making tea" and 600,000

Inference Systems on Object-Use Data

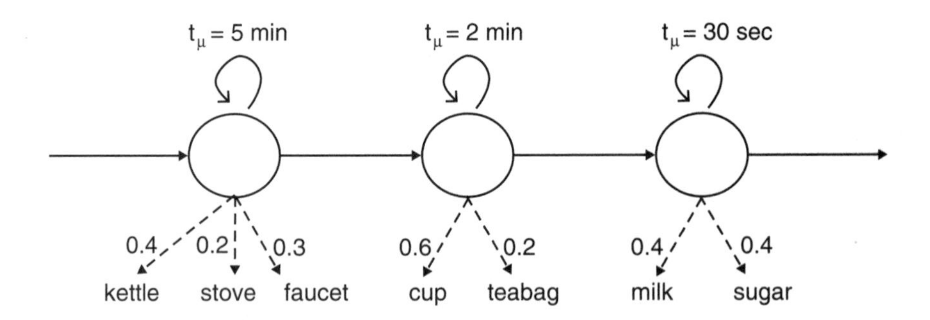

FIGURE 6 A simple probabilistic model for making tea.

of them mention "faucet," these systems accept 60 percent as the rough probability that a faucet is used when making tea. These crude "common-sense" models can be used as a basis for building customized models for each person by applying machine-learning techniques to data generated by that person. Experiments on the data set just described have shown that these completely automatically learned models can recognize activities correctly roughly 50 percent of the time. Analyses of these corpus-based techniques have also provided indirect evidence that object-based models should be sufficient to discriminate between thousands of activities.

CONCLUSIONS

Monitoring day-to-day physical activity is a tedious and expensive task now performed by human monitors. Automated monitoring has the potential of improving the lives of many people, both monitors and those being monitored. Traditional approaches to activity recognition have not been successful at monitoring large numbers of day-to-day activities in unstructured environments, partly because they were unable to identify reliably sufficiently discriminative high-level features. A new family of sensors, based on RFID, is able to identify most of the objects used in activities simply and accurately, and even simple statistical models can classify large numbers of activities with reasonable accuracy. In addition, these models are simple enough that they can extract automatically from massive text corpora, such as the Web, and can be customized for observed data.

ACKNOWLEDGMENTS

This paper describes work done by the author jointly with the SHARP group at Intel Research Seattle and with researchers at the University of Washington. Specifically, work on the iBracelet was done with Adam Rea and Ken Fishkin. The work on WISPs was done with Joshua Smith and the WISP team. The work on mining models was done with Mike Perkowitz, Danny Wyatt, and Tanzeem Choudhury. Inference techniques were developed jointly with Dieter Fox, Henry Kautz, and Don Patterson.

REFERENCES

Finkenzeller, K. 2003. RFID Handbook, 2nd ed. New York: John Wiley & Sons.

Fishkin, K., M. Philipose, and A. Rea. Forthcoming. Hands-On RFID: Wireless Wearables for Detecting Use of Objects. Proceedings of the International Symposium on Wearable Computing, Osaka, Japan, October 18–20, 2005. Washington, D.C.: IEEE Computer Society.

Philipose, M., J. Smith, B. Jiang, A. Mamishev, S. Roy, and K. Sundara-Rajan. 2005. Battery-free wireless identification and sensing. Pervasive Computing 4(1): 37–45.

Philipose, M., K.P. Fishkin, M. Perkowitz, D.J. Patterson, D. Fox, H. Kautz, and D. Hähnel. 2004. Inferring activities from interactions with objects. Pervasive Computing (October-December):50–57. Available online at: *http://seattleweb.intel-research.net/people/matthai/pubs/pervasive_sharp_04.pdf.*

Sanders, B.C.S., R.C. Nelson, and R Sukthankar. 2002. The OD Theory of TOD: The Use and Limits of Temporal Information for Object Discovery. Pp. 777–784 in Proceedings of the Eighteenth National Conference on Artificial Intelligence. Menlo Park, Calif.: AAAI Press.

Tapia, E.M., S.S. Intille, and K. Larson. 2004. Activity Recognition in the Home Setting Using Simple and Ubiquitous Sensors. Pp. 158–175 in Proceedings of PERVASIVE 2004, vol. LNCS 3001, edited by A. Ferscha and F. Mattern. Berlin/Heidelberg: Springer-Verlag. Available online at: *http://courses.media.mit.edu/2004fall/mas622j/04.projects/home/TapiaIntilleLarson04.pdf.*

Wyatt, D., M. Philipose, and T. Choudhury. 2005. Unsupervised Activity Recognition Using Automatically Mined Common Sense. Pp. 21–27 in Proceedings of the Twentieth National Conference on Artificial Intelligence and the Seventeenth Innovative Applications of Artificial Intelligence Conference. Menlo Park, Calif.: AAAI Press.

ENGINEERING FOR DEVELOPING COMMUNITIES

Introduction

Garrick E. Louis
University of Virginia
Charlottesville, Virginia

Amy Smith
Massachusetts Institute of Technology
Cambridge, Massachusetts

Alleviating poverty and ensuring sustainability present daunting challenges and great opportunities for engineering in the 21st century. The challenges involve not only the technical complexities of developing sustainable, affordable, small-scale infrastructure and other engineered systems, but also the "engineering" of difficult interfaces between these technical components and the socioeconomic, cultural, and political contexts in which they are embedded. The opportunities are principally the "opportunity cost" to society of resolving these problems and the size of the potential markets that will benefit from innovations, including 3 billion people who live on less than $2 per day, 2.1 billion who do not have access to sanitation services, and 1.2 billion who do not have access to safe drinking water.

The goal of the session on engineering for developing communities is to facilitate the transfer of knowledge about new techniques and approaches across engineering disciplines to address these problems. This session has three objectives: (1) to inform attendees of the challenges and constraints of engineering for developing communities; (2) to generate discussion and ideas for innovations; and (3) to create a network among attendees for future collaborations.

The presentations focus on three main areas: (1) meeting basic human needs—energy, water, security, health; priorities; challenges; and innovations in technology, methods, and policy; (2) innovations for development—creating added value, opportunities, appropriate technology, indigenous knowledge, role of markets in innovations; and (3) sustainability in a global context—life-cycle analysis, risk analysis, and cost-benefit analysis; policies for energy, security, and ethics; green design. The presentations cover both practical and policy aspects of these issues.

Challenges in the Implementation of Appropriate Technology Projects: The Case of the DISACARE Wheelchair Center in Zambia

KURT L. KORNBLUTH
University of California, Davis

PHILIP OSAFO-KWAAKO
Zambian Ministry of Commerce
Lusaka, Zambia

> ...the most exciting part about these new features is that they do not cost more and we can still deliver a wheelchair for $41.17. Isn't it astounding that God would lead us to a new manufacturer with an innovative approach to our design needs?
>
> *From the Free Wheelchair Mission's Website, June 2005*

The statement above summarizes a major challenge faced by DISACARE[1] Wheelchair Centre in Zambia, an organization that fabricates various models of wheelchairs. At a retail price of US$280, DISACARE makes only a slim profit above net costs. DISACARE provides an interesting example of the implementation of an appropriate-technology project that attempts to use locally-available resources to address technological needs.

Since its modest beginning in 1991, DISACARE has grown from a two-man team to an organization with 23 employees, many of them disabled, who now have expertise in wheelchair fitting and manufacturing, machining, training, accounting, carpentry, and sewing and tailoring. DISACARE is a local Zambian nongovernmental organization (NGO) that has proven itself fiscally responsible and maintained excellent financial reporting to its donors. The organization

[1]All the data cited in this paper were collected by the authors from DISACARE.

provides employment for people with disabilities, as well as advocacy and sports programs. DISACARE's mobility-aid devices are durable and well suited to Zambian conditions, and donors have heralded the organization as a model development project that could be replicated in other economic sectors and other regions. Nevertheless, DISACARE faces some serious challenges.

As a starting point for a discussion of the value of DISACARE as a social-investment scheme, we assess the organization's net benefits, taking into account non-market objectives, such as local skills training and capacity building and poverty-reduction among the disabled. We focus on several key questions. First, was DISACARE a viable development project? Second, was DISACARE a good investment from the point of view of donors? Third, did the intended beneficiaries (the disabled in Zambia) benefit from the DISACARE project?

DISABILITY AID IN ZAMBIA

After independence, most African states pursued state-led industrialization strategies aimed at strengthening local industries and spurring economic growth. In most cases, "infant, import-substitution industries," established under highly protectionist economic policies, were inefficient, resulting in large fiscal deficits and rising debt levels. By the 1990s, most developing countries in Africa were in economic decline. Among the adjustment schemes proposed by international donors were a number of market-liberalization programs intended to stimulate general economic growth (Thirlwall, 2003). By the end of the 1990s, however, the anticipated economic growth in most countries had not been achieved. In fact, levels of poverty and inequality had increased in most communities. In addition, most southern African states were subject to an additional drain on government revenues because of the growing prevalence of HIV/AIDs, with donor policies shifting accordingly (Saasa, 2002).

As a least-developed country (LDC) in southern Africa, Zambia, where support for particularly disadvantaged groups is provided mostly by informal family networks and NGOs, remains highly dependent on the international donor community for support of social-investment schemes. In fact, for most sub-Saharan African countries, official national disability policies are either nonexistent or have not been implemented. In a survey of national disability policies in 83 countries (industrialized, middle-income, and developing countries), the United Nations (UN) Social Commission for Social Development observed widespread gaps in support provided to persons with disabilities in developing countries. For example, in Zambia, Michailakis (1997) reports the absence of regulations for making public buildings accessible to people with disabilities and a lack of coordination and dialogue on disability issues.

DISACARE WHEELCHAIR CENTER

In 1991, Zambia was among the poorest countries in southern Africa, and people disabled by polio had no wheelchairs or prospects for mobility devices. David Mukwasa, grandson of then-Zambian president Keneth Kaunda, and Felix Sulimba, a medical student, both polio victims, envisioned DISACARE in the early 1990s. The two approached the Finish International Disabled Development Association (FIDIDA), and over the next decade, FIDIDA's partner organization, Finish Association of People with Mobility Disabilities (FMD), which played a large role in shaping and supporting DISACARE. They also approached Kenny Mubuyaeta, a polio victim trained in metal fabrication, who left his job in the Copper Belt in northern Zambia and moved to the capital, Lusaka, to join DISACARE and take up the challenge of building wheelchairs.

In a small rented garage, DISACARE began by repairing shopping carts and wheelchairs. Kenny and David, working often without pay, continued trying to get business and raise money. In 1991, Keneth Kaunda's government gave DISACARE a large plot of land in Libala, on the outskirts of Lusaka. In 1995, money was raised from international NGOs for construction of a workshop and small dormitory-style living quarters. In 1996, the workshop in Libala was officially opened. Since 1991, Kenny and David have convinced many local and international organizations to help DISACARE.

Products

DISACARE currently provides a number of locally built mobility-aid products, including: the Kavuluvulu, a standard folding wheelchair outfitted especially for Africa; a tricycle wheelchair, which is specially designed for long distance journeys; a cerebral palsy (CP) wheelchair, which has cushions and tray tables for CP users; and a sports wheelchair, which is specially designed for outdoor sports, such as basketball. DISACARE prices cover all costs, including labor, materials, overhead, and capitol depreciation.

Disability Advocacy

DISACARE currently operates as a trust governed by a board of trustees with the following mission: "To provide mobility, empowerment, and self-sustenance for persons with disabilities." DISACARE is the only domestic manufacturer of wheelchairs; the company also repairs and customizes wheelchairs. In addition, DISACARE has been a major advocate for people with disabilities. The organization provides appropriate mobility aids, as well as training and employment. DISACARE has greatly increased the visibility of people with disabilities in the community. Both Kenny and David have polio but drive cars, and DISACARE workers have broken the stereotype that disabled people are unable to work and are simply beggars. Twice a week, wheelchair riders gather at DISACARE to play wheelchair basketball.

Sales of Wheelchairs

DISACARE began repairing wheelchairs in 1991 and gradually started production in 1993. As a result of poor marketing, high overhead costs, and weak purchasing power in the domestic market, however, sales reached a peak of 200 in 2003, but have gone down since then. DISACARE must sell between 180 and 240 wheelchairs per year to pay salaries. However, to be self-sustaining, the company must sell more than that.

TABLE 1 Wheelchair Sales and Repairs

Year	Wheelchairs Produced	Wheelchairs Repaired
1991	—	15
1992	—	200
1993	5	120
1994	7	100
1995	8	60
1996	12	60
1997	20	80
1998	50	100
1999	70	60
2000	80	50
2001	115	40
2002	190	50
2003	197	40
2004	100 (Aug)	20
TOTAL	854	995

Source: DISACARE, 2004.

The Competition

Wheelchairs produced in Zambia face stiff competition from wheelchairs imported from Asia. In addition, DISACARE has high overhead costs and, because it uses local materials, it does not take advantage of cheaper imported raw materials. The basic DISACARE wheelchair costs $280, is locally repairable, and lasts from two to five years, with proper maintenance. Although this seems inexpensive compared to a typical folding wheelchair, which costs $500 to $1,000 in the United States or Europe, many aid organizations prefer to donate cheaper Chinese-made wheelchairs, although they often have poor ergonomics, are not suitable for local conditions, and have a short life (six months to two years). Nevertheless, many donors have opted to buy them because, at $75 to $200 apiece, they can provide them to more clients for the same cost.

Providing Employment

Providing employment is an important objective of DISACARE. As DISACARE has expanded its operations in the past decade, it has increased the size of its workforce to 23, boosting employment in the local economy. DISACARE is an affirmative-action employer, and more than half of its workers have physical disabilities.

Capacity Building

Donors and volunteers have built the capacity of DISACARE by supplying experts to train staff members in wheelchair design, accounting, fundraising and administration, and sports. As a result, the DISACARE staff not only understands the needs of people with disabilities, but also is adept at running day-to-day operations.

TABLE 2 Capacity Building

Year	Specialty	Comments
1991-1997	Prothetist	FMD volunteer
1993	Wheelchair designer	WWI paid consultant
1997	Wheelchair sports	FMD paid consultant
1997-2001	Wheelchair designer	WWI paid consultant
1997-present	Wheelchair designer	WWI paid consultant
2001, 2004	Product designer	DEKA volunteer
1999-2001	Administration/Fundraising	VSO volunteer
2002	Administration/Fundraising	VSO volunteer
2002	Accounting	VSO volunteer

Source: DISACARE, 2004.

Financing of DISACARE

Throughout its short history, DISACARE has relied heavily on external donor support for financing and capitalization (Table 3).

Current Assets

Much of the donor investment in DISACARE, such as the buildings and land near Lusaka, is still owned by the DISACARE trust. The estimated current value of DISACARE's assets is shown in Table 4.

TABLE 3 Sources of Funding

Year	Donor	Monetary value	Description
1991	FIDIDA	$80	Tools
1991	Meal-a-Day	$3,500	Tools, equipment
1991	Dutch Embassy	$3,000	Equipment and materials
1992	Gov't of Zambia	$200	3.5 acre plot
1992-1997	FMD	$10,000 (est.)	Equipment, materials, machinery
1993	FMD	$15,000	WWI training
1995	Beit Trust	$25,000	Accommodation blocks
1997	FMD	$1,000	Basketball court
1999	Abillis	$5,000	Office equipment, materials
1999	British Embassy	$7,000	Office equipment, materials, building
2000	German Embassy	$9,000	Lathe
2001	Irish Aid	$11,000	Milling machine
2001	Danish Embassy	$8,000	Materials, equipment
2001	Beit Trust	$6,000	Workshop extension
2001	FMD	$4,500	Vehicles
2002	US AID	$5,000	Office equipment, furniture
1998-2002	FMD	$60,000 (est.)	Wheeling wheels project
2003	FMD	$8,000	Capital, building, shop upgrade
2003	Barclays Bank	$400	Machine tools
2004	FMD	$1,000	Vehicles
TOTAL		$182,680	

Source: DISACARE , 2004.

TABLE 4 Current Assets, December 2003

Item	Value
Land	$25,000
Buildings	$80,000
Sports toilets	$2,500
Basketball court	$2,000
Inventory	$12,000
Containers	$4,000
Machinery	$30,000
Office equipment	$10,000
Cash	$7,000
Furniture	$2,000
Vehicles	$7,500
TOTAL	$182,000

Source: DISACARE, 2004.

SUMMARY

Whereas many donor investment projects end up bankrupt after five or ten years, the current assets of DISACARE are roughly equal to the total donor investment. A brief summary of the benefits and costs follows.

Costs to Donor Community

- Total cash investment equals $182,000.
- In-kind services are provided by volunteers.

Benefits to Zambian Disabled Community

- More than 1,750 disabled people have been provided with wheelchairs.
- Currently, 26 Zambians are employed.
- Industrial capacity in Zambia has increased.
- Capacity in the eastern/southern African region has increased.
- Awareness of people with disabilities in Zambia has increased.

Including repaired wheelchairs, DISACARE's has provided mobility for 1,750 people with disabilities in its 14-year lifetime. This translates to $182,000/1,750 people, or a little more than $100 per person.

REFERENCES

Michailakis, D. 1997. Government Action on Disability Policy: A Global Survey. Stockholm: Liber Publishing House.

Saasa, O. 2002. Aid and Poverty Reduction in Zambia: Mission Unaccomplished. Uppsala, Sweden: Nordic Africa Institute.

Thirlwall, A.P. 2003. Growth and Development: With Special Reference to Developing Countries. London: Macmillan.

Engineering Inputs to the CDC Safe Water System Program

DANIELE S. LANTAGNE
U.S. Centers for Disease Control and Prevention
Atlanta, Georgia

In September 2000, the United Nations General Assembly adopted the Millennium Development Goals (MDGs) to promote "human development as the key to sustaining social and economic progress" (World Bank Group, 2004). One MDG target is to "halve, by 2015, the proportion of people without sustainable access to safe drinking water (1.1 billion) and basic sanitation (2 billion)" (World Bank Group, 2004). Although the world is on schedule to meet the water-supply target (WHO/UNICEF, 2004), even if the goal is met, more than 600 million people will still lack access to safe water in 2015 (WHO/UNICEF, 2000). In addition, although the MDG target specifically states "safe" drinking water, assessing the safety of water at the household level is difficult; thus, the metric for assessing the target is access to water from "improved" sources, such as boreholes or household connections (WHO/UNICEF, 2004). Therefore, people who drink unsafe water from improved sources will not be impacted by the MDG.

The health consequences of inadequate water and sanitation services include an estimated 4 billion cases of diarrhea and 2.2 million deaths each year, mostly among young children in developing countries (WHO/UNICEF, 2000). In addition, waterborne diarrheal diseases lead to decreased food intake and nutrient absorption, malnutrition, reduced resistance to infection (Baqui et al., 1993), and impaired physical growth and cognitive development (Guerrant et al., 1999).

THE SAFE WATER SYSTEM PROGRAM

Chlorination, first used for disinfection of public water supplies in the early 1900s, has drastically reduced waterborne disease in cities in the developed world (Cutler and Miller, 2005). Elsewhere, although small trials of point-of-use chlorination have been conducted in the past (Mintz et al., 1995), large-scale trials began only in the 1990s, as part of the Pan American Health Organization (PAHO) and U.S. Centers for Disease Control and Prevention (CDC) response to epidemic cholera in Latin America (Tauxe, 1995). The Safe Water System (SWS) Program devised by CDC and PAHO includes three elements: (1) water treatment with dilute sodium hypochlorite at the point of use; (2) storage of water in a safe container; and (3) education to improve hygiene and water practices. In four randomized, controlled trials, SWS resulted in a 44 to 84 percent reduction in the risk of diarrheal disease (Luby et al., 2004; Quick et al., 1999, 2002; Semenza et al., 1998).

Subsequent SWS implementation has varied according to local partnerships and social and economic conditions. The disinfectant solution has been disseminated in 13 countries at national and subnational levels through social marketing in partnership with Population Services International (PSI), a nongovernmental organization (NGO). In Ecuador, Laos, Haiti, and Nepal, the Ministry of Health or a local NGO have implemented the SWS Program at the community level. SWS has also been made available free of charge following disasters in a number of places, including Indonesia, India, and Myanmar (following the 2004 tsunami) and Kenya, Bolivia, Haiti, Indonesia, and Madagascar (after other natural disasters).

Engineering Input: Standardized Methodology, Dose Factor, and Regionalization

The instructions on the bottle of sodium-hypochlorite solution ("product") used by SWS advises users to add one full cap of the solution to clear water (or two caps to turbid water) in a standard-sized storage container, agitate, and wait 30 minutes before drinking. From 1993 to 2003, CDC assisted PSI in eight countries, and NGOs or government ministries in six additional countries, to establish SWS projects. In each location, the best packaging option for the solution was either selected from available existing plastic bottles and caps or contracted in-country for new bottle and cap designs. A dosing strategy and sodium-hypochlorite concentration were selected after testing chlorine dosages. This strategy led to wide variations in products, with some caps 22-millimeters in diameter with a volume of 8 to 10 mL (Figure 1). For these large-volume caps, the concentration of sodium hypochlorite had to be lowered to ensure correct dosing.

The use of widely varying products makes it difficult to compare dosing

FIGURE 1 Initial SWS products (Bolivia, Peru, Zambia, Uganda, Kenya, India, Madagascar). Source: U.S. Centers for Disease Control and Prevention.

strategies. A loading factor, called the "dose factor," was developed for comparing dosing strategies for different packaging options. The dose factor is simply the hypochlorite concentration in percent multiplied by the amount (in mL) of hypochlorite solution added to 20 liters of clear water for treatment. The unit of measurement is %-mL, which is convenient for measuring water use (Equation 1). However, scientifically, the dose factor is the concentration of sodium hypochlorite added to 20 liters of water (Equation 2).

$$\text{Dose Factor } (\% - \text{mL}) = \text{Hypochlorite Concentration } (\%) \times \text{Amount added to 20 liters (mL)} \quad (1)$$

$$\text{Dose } (\text{mg}/\text{L}_\text{w}) = \frac{\text{Hypochlorite Concentration (mc/Lcl)} \times \text{Amount added (mLcl)} \times 1\ \text{L}/1000\text{mL}}{20\ (\text{Lw})} \quad (2)$$

There was a large range of dose factors in the initial SWS products (1.6 to 8.0, median 4). To determine the cause of this variation, a consistent methodology was developed to complete dosage testing for SWS implementation. This methodology has since been completed in 15 countries. The methodology includes: (1) collecting three 20-liter containers of water from six representative sources; (2) testing the water for turbidity, pH, conductivity, and residual chlorine; (3) adding a dose factor of 1.875, 3.75, and 7.5 of sodium hypochlorite to one container from each source; and (4) testing free and total residual chlorine at 1, 2, 4, 8, and 24 hours after the addition of chlorine to determine the dose factor that leads to a free chlorine residual value of less than 2.0 mg/L one hour after the addition and greater than 0.2 mg/L 24 hours after the addition. The one hour value of free chlorine residual (2.0 mg/L) represents the limit of user acceptability, and the 24 hour value represents the minimum level for protection from recontamination. This range also meets the World Health Organization (WHO) guidelines for free chlorine in drinking water (WHO, 2005).

In 73 (84 percent) of 87 samples of unchlorinated water from 13 countries, a dose factor of 3.75 for "clear-appearing" water, and 7.5 for "dirty-appearing water," was appropriate. The remaining 14 (16 percent) samples had excessive turbidity (57 percent), metals leading to chlorine demand (21 percent), or were best treated with a dose factor of either 1.875, or between 3.75 and 7.5 (21 percent).

In Angola, 11 of 12 water tests fell into the 3.75 to 7.5 dose factor regime (Figure 2). Water in the Luanda water treatment plant (the tanker-truck filling

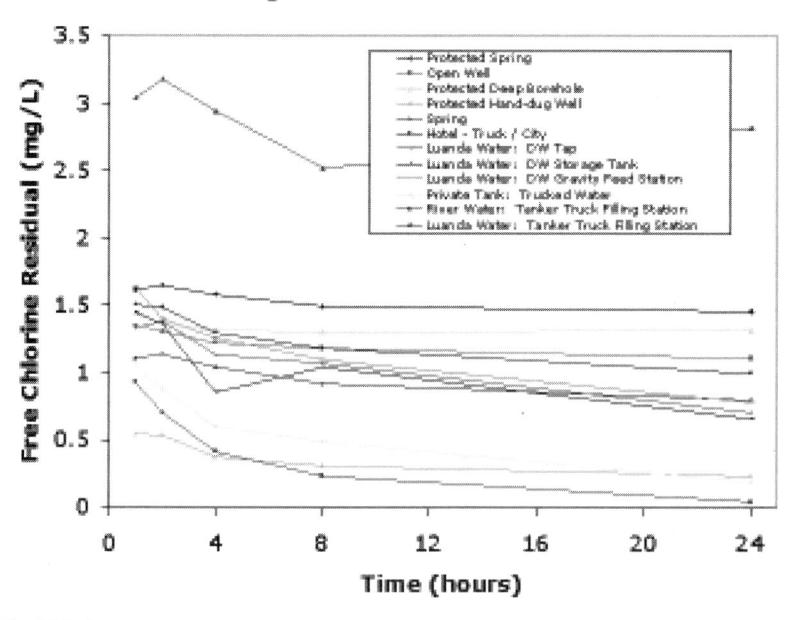

FIGURE 2 Dosage testing results in Angola. Source: U.S. Centers for Disease Control and Prevention.

station) was already chlorinated and thus contained more than 2.0 mg/L free chlorine at all times; the river water had higher chlorine demand, necessitating a dose factor of 7.5.

These consistent results are in contrast to the wide dosage range in the initial products. Thus, the wide initial range can be attributed to compromises in dosing to fit available bottles, inconsistent testing methodologies, and the testing of unrepresentative water sources in pilot project areas.

Based on the consistent results in Angola, we were able to design a standard product. After analyzing design variables, dosing requirements, and constraints, such as transport considerations and available label size, a 150-mL bottle with a 3-mL cap and 1.25 percent hypochlorite solution was chosen. Adding one cap per 20 liters of water (dose factor 3.75), this bottle can treat 1,000 liters of clear water and last a family of 5 or 6 for approximately one month.

PSI (the social-marketing NGO that operates in 70 countries) is adopting this ideal product regionally. PSI designs brand names for health products, sells them at low cost, distributes them through wholesale and retail commercial networks, and generates demand through behavior-change communications, such as radio and TV spots, mobile video units, point-of-sale materials, theatrical presentations, and person-to-person communication. Currently, PSI has SWS programs in 13 countries; programs in eight additional countries are expected to be launched by mid-2006. PSI sold more than 12 million bottles of SWS solution in 2004, and has adopted or will adopt the regional product in 13 countries, including Kenya (Figure 3).

PSI implements the regional product by importing a bottle mold (US$2,200) for installation in a local plastics company. In Nairobi, Kenya, US$12,000 was invested in the design and manufacture of a 3-mL volume cap mold (by AMM Engineering) and production of the caps (by BlowPlast Industries) for just

FIGURE 3 PSI Kenya regional product. Source: PSI Kenya.

US$0.011 each. The caps, which are difficult to produce but easy to transport, are then shipped to other countries in the region. Regional production has significantly reduced product cost (e.g., by 55 percent in Madagascar), simplified project initiation, and facilitated the importation of product from neighboring countries in emergencies. The SWS program operated by PSI has not only had an impact at the country level but has also brought these benefits to regions throughout Africa and Asia.

CONCLUSION

The SWS is a proven, low-cost intervention that has the potential of providing safe drinking water in areas where there will be no infrastructure to treat water in the near future and of significantly reducing morbidity from waterborne diseases and improving the quality of life for millions of people. As SWS was scaled up from pilot projects to at-scale implementation, engineering inputs were critical to ensuring that effective, high-quality, easy-to-use, cost-effective products were designed and produced. To achieve, and even surpass, the MDG, will require continued collaboration between engineers, public health professionals, and in-country implementers. SWS and similar point-of-use treatment programs are our best hope for rapid reductions in the burden of waterborne diseases and deaths in developing countries.

REFERENCES

Baqui, A.H., R.E. Black, R.B. Sack, H.R. Chowdhury, M. Yunus, and A.K. Siddique. 1993. Malnutrition, cell-mediated immune deficiency and diarrhea: a community-based longitudinal study in rural Bangladeshi children. American Journal of Epidemiology 137(3): 355–365.

Cutler, D., and G. Miller. 2005. The role of public health improvements in health advances: the twentieth-century United States. Demography 42(1): 1–22.

Guerrant, D.I., S.F. Moore, A.A. Lima, P.D. Patrick, J.B. Schorling, and R.L. Guerrant. 1999. Association of early childhood diarrhea and cryptosporidiosis with impaired physical fitness and cognitive function seven years later in a poor urban community in northeast Brazil. American Journal of Tropical Medicine and Hygiene 61(5): 707–713.

Luby, S.P., M. Agboatwalla, R. Hoekstra, M. Rahbar, W. Billhimer, and B. Keswick. 2004. Delayed effectiveness of home-based interventions in reducing childhood diarrhea, Karachi, Pakistan. American Journal of Tropical Medicine and Hygiene 71(4): 420–427.

Mintz, E., F. Reiff, and R. Tauxe. 1995. Safe water treatment and storage in the home: a practical new strategy to prevent waterborne disease. Journal of the American Medical Association 273(12): 948–953.

PSI (Population Services International). 2005. Population Services International. Available online at: *http://www.psi.org* (June 23, 2005).

Quick, R., L. Venczel, E. Mintz, L. Soleto, J. Aparicio, M. Gironaz, L. Hutwagner, K. Greene, C. Bopp, K. Maloney, D. Chavez, M. Sobsey, and R. Tauxe. 1999. Diarrhea prevention in Bolivia through point-of-use disinfection and safe storage: a promising new strategy. Epidemiology and Infection 122(1): 83–90.

Quick, R.E., A. Kimura, A. Thevos, M. Tembo, I. Shamputa, L. Hutwagner, and E. Mintz. 2002. Diarrhea prevention through household-level water disinfection and safe storage in Zambia. American Journal of Tropical Medicine and Hygiene 66(5): 584–589.

Semenza, J., L. Roberts, A. Henderson, J. Bogan, and C. Rubin. 1998. Water distribution system and diarrheal disease transmission: a case study in Uzbekistan. American Journal of Tropical Medicine and Hygiene 59(6): 941–946.

Tauxe, R. 1995. Epidemic cholera in the New World: translating field epidemiology into new prevention strategies. Emerging Infectious Diseases 1(4): 141–146.

WHO (World Health Organization). 2005. Letter addressed to Dr. Eric Mintz stating SWS dosing regime complies with WHO guideline values. Email dul4@cdc.gov for PDF file.

WHO/UNICEF (United Nations Children's Fund). 2000. Global Water Supply and Sanitation Assessment, 2000 Report. Geneva: Water Supply and Sanitation Collaborative Council, WHO. Available online at: *http://www.who.int/water_sanitation_health/monitoring/globalassess/en/.*

WHO/UNICEF. 2004. Meeting the MDG Drinking Water and Sanitation Target: A Mid-Term Assessment of Progress. Available online at: *http://www.who.int/water_sanitation_health/monitoring/en/jmp04.pdf.*

World Bank Group. 2004. Millenium Development Goals: About the Goals. Available online at: *http://www.developmentgoals.org/About_the_goals.htm* (September 2004).

Sustainable Development Through the Principles of Green Engineering

Julie Beth Zimmerman
University of Virginia
Charlottesville, Virginia

As concerns about population growth, global warming, resource scarcity, globalization, and environmental degradation have increased, it has become apparent that engineering design must be engaged more effectively to advance the goal of sustainability. This will require a new design framework that incorporates sustainability factors as explicit performance criteria. Sustainability has been defined as "meeting the needs of the current generation without impacting the needs of future generations to meet their own needs" and is often interpreted as the simultaneous advancement of prosperity, environment, and society. The 12 Principles of Green Engineering developed by Anastas and Zimmerman (2003) provide a design protocol for moving toward engineering design for sustainability.

The impact of population growth has long been understood as a grand challenge to the advancement of sustainability goals. The data demonstrate that the vast majority of population growth is occurring in the developing world and that population is stagnant, in some cases declining, in the industrialized world (Figure 1). This may suggest that in the complex equation of a growing world population, including birth and mortality rates, socio-political pressures, access to health care and education, cultural norms, and so on, there is an empirical correlation between the rate of population growth and the level of economic development, often equated with quality of life.

This relationship suggests that one approach to meeting the challenges of stabilizing population growth and advancing the goals of sustainability is to

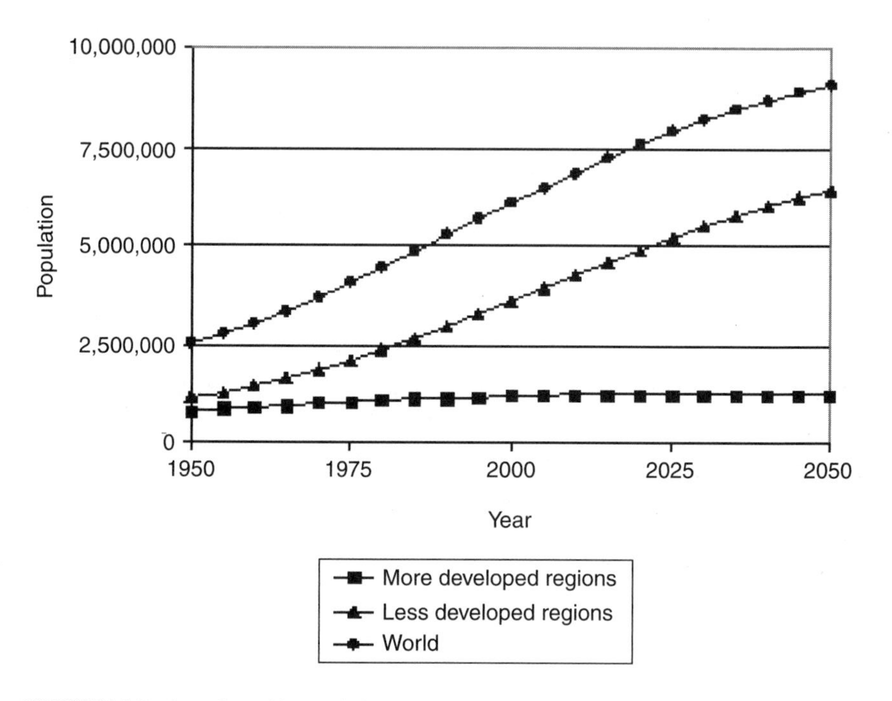

FIGURE 1 Projected world population. More developed regions comprise all of Europe, North America, Australia/New Zealand, and Japan. Less developed regions comprise all of Africa, Asia (excluding Japan), Latin America and the Caribbean, Melanesia, Micronesia, and Polynesia. Source: UNDESA, 2004.

expand economic development and improve quality of life. Historically, however, increased development and higher quality of life have been inextricably linked with environmental degradation and resource depletion. A significant amount of evidence suggests that the growing human population has increased the strain on natural resources used for consumption and waste assimilation.

Although there is no single index of the state of the environment, the relationship between population and environment can be analyzed in terms of resource depletion or dimensions of environmental quality, such as land use, water quantity and quality, pollution (particularly from energy demand), biodiversity, and climate change. A brief review of each of these indicators supports the notion that population growth has traditionally had a detrimental impact on the environment.

The question, therefore, is how to bring about continued development and improved quality of life in both the developing and developed world without environmental degradation and excessive resource consumption. The goal of green engineering and green chemistry is to use science and technology to ensure that quality of life, or economic development, is improved through benign chemicals and materials and life-cycle-based design, as well as material and energy efficiency and effectiveness (Anastas and Warner, 1998). Achieving this goal would decouple the historical links between improved quality of life, population growth, and environmental degradation.

THE 12 PRINCIPLES OF GREEN ENGINEERING

The 12 Principles of Green Engineering (see Table 1) provide a framework for designing new materials, products, processes, and systems that are benign to human health and the environment. A design based on the 12 Principles goes beyond baseline engineering quality and safety specifications to sustainability factors, which are considered fundamental factors from the earliest stages of design of a material, product, process, building, or a system. These principles were developed to frame design architecture—whether molecular architecture to construct chemical compounds, product architecture to create an automobile, or urban architecture to build a city. The 12 Principles are applicable, effective, and appropriate for all of them. To function as global design principles, they must be independent of local parameters and system conditions.

TABLE 1 The 12 Principles of Green Engineering

Principle 1. Designers should ensure that all material and energy inputs and outputs are as inherently non-hazardous as possible.
Principle 2. It is better to prevent waste than to treat it or clean it up after it is formed.
Principle 3. Separation and purification operations should be a component of the design framework.
Principle 4. System components should be designed to maximize mass, energy, and temporal efficiency.
Principle 5. System components should be output-pulled rather than input-pushed through the use of energy and materials.
Principle 6. When making design choices on recycling, reuse, and beneficial disposition, embedded entropy and complexity should be considered an investment.
Principle 7. Targeted durability, not immortality, should be a design goal.
Principle 8. Design for unnecessary capacity or capability should be considered a design flaw. This includes engineering "one size fits all" solutions.
Principle 9. Multi-component products should strive for material unification (minimal material diversity) to promote disassembly and value retention.
Principle 10. Design of processes and systems must include integration of interconnectivity with available energy and materials flows.
Principle 11. Performance metrics should include designing for performance in commercial "afterlife."
Principle 12. Design should be based on renewable and readily available inputs throughout the life cycle.

Source: Anastas and Zimmerman, 2003.

Two fundamental concepts engineers should strive to integrate at every opportunity are (1) life-cycle considerations and (2) inherency. The materials and energy that enter each life cycle stage of every product and process have their own life cycles. If a product is environmentally benign but requires hazardous or nonrenewable substances to produce, the environmental impacts have simply been shifted to another stage in the overall life cycle. Thus, designers must consider the entire life cycle, including the life cycles of materials and energy inputs. This strategy complements the selection of inherently benign inputs that will reduce environmental impacts throughout the life cycle.

Making products, processes, and systems more environmentally benign generally follows one of the two basic approaches: (1) changing the inherent nature of the system; or (2) changing the circumstances/conditions of the system. For example, inherency may reduce the intrinsic toxicity of a chemical. A conditional change might be to control the release of, and exposure to, a toxic chemical. Inherency is preferable for various reasons, most importantly because it precludes "failure." In the example just described, technological control of system conditions could potentially fail, which could lead to a significant risk to human health and natural systems. With an inherently benign design, regardless of changes in conditions or circumstances, the intrinsic nature of the system cannot fail.

The 12 Principles provide a structure for creating and assessing the elements of design to maximize sustainability. The application of the 12 Principles on different scales and in different disciplines has been documented with case studies from a variety of sectors (Zimmerman and Anastas, 2005; Zimmerman et al., 2003). Although designers of molecular systems, designers of industrialized systems, and designers of systems for developing communities use different terminology and jargon, the fundamental approaches and guidelines apply to all of them. The case studies illustrate how the framework of principles has worked in the past and provide a blueprint for applying them in future designs for improving quality of life and ultimately advancing sustainability.

ADVANCING GLOBAL SUSTAINABILITY

Science and technology are vital to advancing global sustainability through the next-generation design of fundamental products, processes, and systems that not only maintain and/or improve quality of life, but also protect the planet. The current operational model of unilateral knowledge transfer from the industrialized world to the developing world could also be expanded to include knowledge exchange (dialogue), which would allow for learning about indigenous knowledge and traditional designs that have been developed and adapted for local people and places. Knowledge exchange would provide an opportunity for integrating the knowledge from different sources and different methodologies, techniques, and practices from the developed and developing worlds. The examples of innovations in science and technology from the developing world would highlight alternative strategies to delivering services, such as clean drinking water, medical treatment, energy and power production, material and product development, and building technologies and techniques.

Developing nations typically have a long history of practical innovation and successful application of indigenous knowledge systems to serve individuals and communities (Mihelcic et al., 2005). Innovations in science and technology can lead the way to fundamental changes in the quantity and types of energy and materials used to improve quality of life and advance prosperity while protecting and restoring natural systems. The incorporation of cutting-edge thinking and traditional ways will create a robust effort to achieve the common goal of sustainable development.

CONCLUSIONS

Achievements based on green engineering principles are examples of how products and systems have been designed with a new sustainability perspective. To address the challenges of sustainability, in both industrialized and, especially, developing nations, where development will be most consequential for the environment and society, new design imperatives must be systematically incorporated into the next generation of products, processes, and systems. In this context, the dialogue between the developed and developing world must include not only a high-level understanding of complex systems, but also the simple elegance of solutions based on millennia of experience and tradition. Designing more sustainable systems and products will require diverse sources of technological inspiration.

REFERENCES

Anastas, P, and J. Warner. 1998. Green Chemistry: Theory and Practice. London: Oxford University Press.

Anastas, P., and J. Zimmerman. 2003. Design through the 12 Principles of Green Engineering. Environmental Science and Technology 37(3): 94A–101A.

Mihelcic, J., A. Ramaswami, and J. Zimmerman. 2005. Integrating Developed and Developing World Knowledge into Global Discussions and Strategies for Sustainability. Submitted to Environmental Science and Technology.

UNDESA (United Nations Department of Economic and Social Affairs). 2004. World Population Projections to 2050. New York: Population Division, United Nations Department of Economic and Social Affairs.

Zimmerman, J.B., and P.T. Anastas. 2005. The 12 Principles of Green Engineering as a Foundation for Sustainability in Sustainability Science and Engineering: Principles, edited by M. Abraham. Amsterdam: Elsevier Science. In press.

Zimmerman, J.B., A.F. Clarens, S.J. Skerlos, and K.F. Hayes. 2003. Design of emulsifier systems for petroleum- and bio-based semi-synthetic metalworking: fluid stability under hardwater conditions. Environmental Science and Technology 37(23): 5278–5288.

Science and Engineering Research That Values the Planet

Daniel M. Kammen
University of California, Berkeley

Arne Jacobson
Humboldt State University
Arcata, California

The recognition that human activity is transforming the planet, both in intended and dramatically unintended ways, has led to the development of a new field of research—sustainability science. Widely discussed essays (e.g., Clarke, 2002; Kates et al., 2001; Kennedy, 2003; McMichael et al., 2003; Swart et al., 2002), special issues of premier journals (NAS, 2003), and extensive websites (FSTS, 2005) are now devoted to defining sustainability and identifying useful modes and topics for research. Building on this foundation, we now have a tremendous opportunity to advance a new global scientific research paradigm—the generation and implementation of sustainability science. One important lesson emerges very clearly from this body of work—only by posing the question of sustainability explicitly and, where necessary, repairing the damage humans have caused to the biosphere, can we begin to understand how humans can prosper without degrading the planet.

In a seminal treatise on science policy, Vannevar Bush (1945) wrote that, "applied research invariably drives out pure [research]," to the detriment, in his view, of the national capacity for innovation. The subsequent separation of basic and applied research shaped the evolution of science and engineering research for decades and was a point of departure for E.F. Schumacher (1973) and the "appropriate technology" movement, a precursor of sustainability science that involved identifying important but neglected issues for scientific study. This approach, dubbed "mundane science," (Kammen and Dove, 1997), involves projects that combine pragmatic and goal-oriented applied research with potential advances in basic science (Stokes, 1997). The growing recognition of the value of supporting interdisciplinary research and the emergence of sustainability science are continuations of the intellectual evolution of the interaction between science and society.

The scientific recognition of the reality of global environmental change (Hansen et al., 2005), the political awareness of the need to act now to address greenhouse gas emissions (Kennedy, 2005), and the increasing disparities between the lives of the poor and the wealthy provide an opportunity for galvanizing global action to place sustainability science at the forefront of educational, research, and career-development agendas. The next step toward putting sustainable science into action is recognizing that, with ecological stewardship as a guiding scientific principle, entirely new avenues of inquiry are possible.

At this moment in history, this message has the potential to transform research careers and make sustainability a theme that researchers, public officials, and civil society can all embrace. The World Conference on Physics and Sustainable Development, held in Durban, South Africa, in October and November 2005, provided a forum for showcasing opportunities for the co-evolution of basic research and social advances (SAIP, 2005).

Currently attention, debate, and a trans-Atlantic division are focused on how to provide meaningful, long-term aid and assistance to Africa. To highlight a potential solution, we present two cases of sustainable science, engineering, and action in developing nations that advance both science and sustainable human and ecological communities.

THE ENERGY-HEALTH-ECOLOGY NEXUS

Household use of solid fuels is one of the leading causes of death and disease in developing countries throughout the world—particularly among women and children (Smith et al., 2004). Over the past decade, a series of studies has been conducted of programs to design and disseminate more efficient, safer household stoves and to develop and implement sustainable forestry and fuel (often charcoal) production practices in Africa. As Figure 1 shows, combined attention to both stove and forestry programs can lead to dramatic simultaneous improvements in human health, ecological sustainability, and local economic development (Kammen, 1995).

The Kenya study showed that transitions from wood and dung fuels burned in simple stoves to charcoal burned in improved stoves reduced the frequency of acute respiratory infections (ARI) by a *factor of two*. This is a tremendous impact on ARI, the most common illnesses reported in medical exams in sub-Saharan Africa. Comparatively simple materials and design modifications to household stoves are now known not only to improve energy efficiency, but also to reduce particulate and greenhouse gas emissions (Bailis et al., 2005).

These benefits can be achieved at exceptionally low cost, just a few dollars

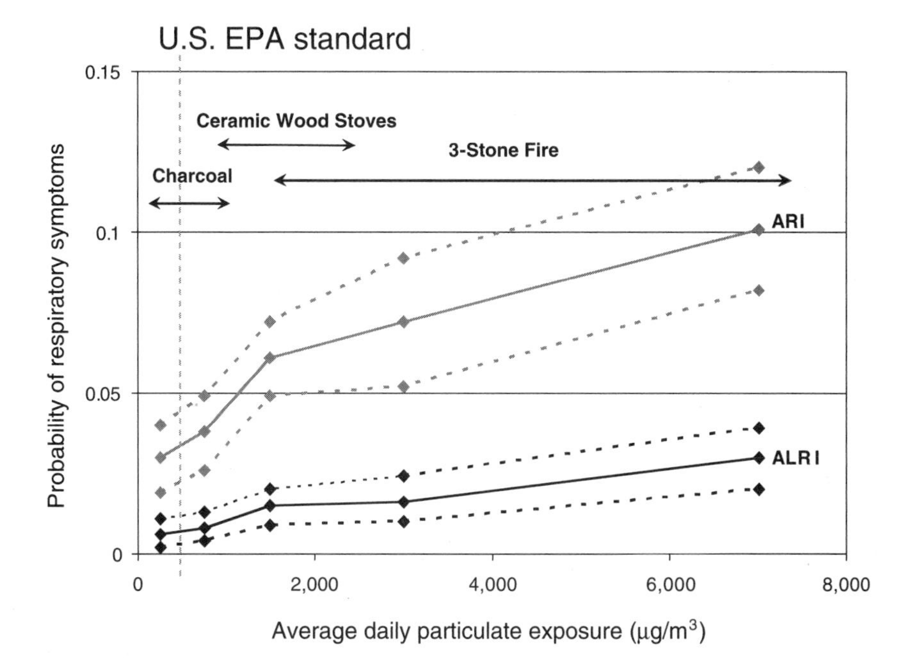

FIGURE 1 The exposure-response graph from a six-year, 500-person exposure and stove intervention study in Kenya. The vertical axis shows the percentage of time subjects participating in biweekly health examinations exhibited ARI or acute lower respiratory illness (ALRI) symptoms. The EPA particulate exposure standard of 200 $\mu g/m^3$ for PM_{10} (particles with diameters of less than 10 microns) is indicated by the dotted vertical line, which forms a lower bound for the exposure range observed in the Kenya project. The stove and fuel combinations indicate exposure ranges. Adapted from Ezzati and Kammen, 2001.

per life saved, and have the added benefit of mitigating atmospheric carbon, at just a few dollars per ton of carbon (Ezzati and Kammen, 2002). By contrast, carbon today trades for roughly $30/ton on the London exchange, a price that reflects only the impact of greenhouse gases. By making the dissemination and use of improved cookstoves a component of a comprehensive Africa-assistance strategy, both local health and development needs *and* global environmental protection could be addressed with great economic efficiency.

The project in Kenya led to a number of unanticipated advances in "basic science." The high pollution concentrations observed in rural African homes—as much as 100 times higher than those observed in the urban areas of many industrialized nations—provided a laboratory for examining the epidemiology of exposure-response in a pollution regime that had not been studied before (Ezzati and Kammen, 2001). These studies have greatly extended the cutting-edge epidemiological work being done largely in developed nations (Rich et al., 2005).

SOLAR ELECTRICITY MARKETS IN DEVELOPING NATIONS

Household solar photovoltaics (PV) have emerged as the leading alternative to grid-based rural electrification in many developing countries. In Kenya, 30,000 PV systems are sold annually, making it a global leader, per capita, in sales of residential renewable energy systems (Figure 2). Advances in amorphous silicon (a-Si) PV technology, which led to the development of small, low-cost a-Si PV modules, played a critical role in the emergence and growth of the Kenyan solar market (Hankins, 2000; Jacobson, 2004).

A key aspect of these advances involved minimizing the initial light-induced Staebler-Wronski degradation of a-Si modules, a poorly understood materials issue with significant implications for low-cost solar cells. The power output of a-Si solar modules typically decreases by 15 to 40 percent during the first few months of exposure to solar radiation due to Staebler-Wronski degradation. Better quality brands have lower degradation levels (Staebler and Wronski, 1977; Su et al., 2002), and after the initial period of degradation, the power output stabilizes. Figure 3 shows degradation curves for two different brands of a-Si modules, showing that the initial power output of some brands drops significantly more than others. The rated power of most reputable brands of a-Si PV

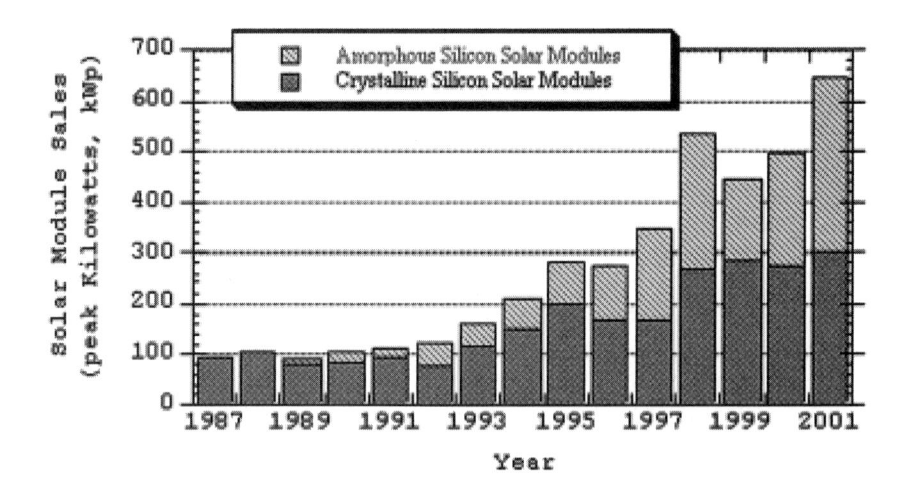

FIGURE 2 Sales of solar modules from 1987 to 2001 in Kenya showing the dramatic increase in sales of amorphous silicon (a-Si) solar cells. The average system size is less than 25 Wp, and current annual sales exceed 30,000 individual solar electric home systems. A substantial fraction of crystalline silicon (c-Si) module sales are for institutional systems that are funded primarily through donor aid programs. Sources: ESDA, 2003; Hankins, 2000; Hankins and Bess, 1994.

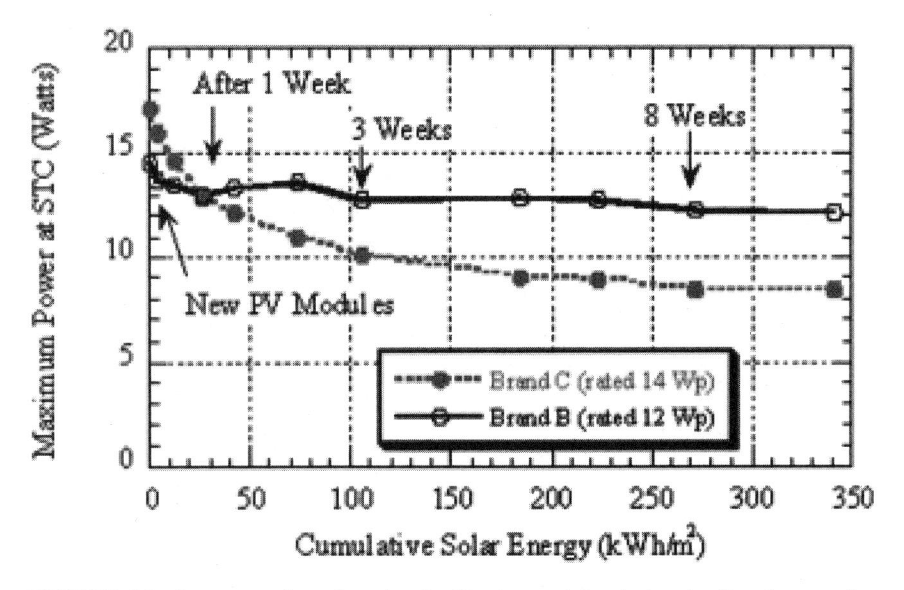

FIGURE 3 Performance of two brands of a-Si solar modules during the first few months of exposure to solar radiation showing substantial differences in light-induced Staebler-Wronski degradation for the two brands. The power output of the Brand C module, although initially higher than its 14W power rating, drops far below its nameplate rating after several months in service. By contrast, the performance of the Brand B module stabilizes near the 12W rating. Note that these results are from 2000 and do not reflect recent improvements for Brand C (shown in Figure 5). Source: Jacobson et al., 2000. Reprinted with permission.

modules corresponds to the final, stabilized power output under standard test conditions of 1,000 W/m^2 and 25°C.

A second important design issue has been the development of cost-effective sealant materials and methods of preventing delamination. Water intrusion can lead to outright module failure, and the actual power output of modules with significant delamination is often reduced to less than 10 percent of the nameplate power rating. Figure 4 shows water-induced delamination in an a-Si module caused by low-quality seals. A number of a-Si manufacturers have developed highly effective sealing techniques, but a few brands continue to have water-intrusion problems.

These advances have been important for the PV industry as a whole, but have been especially significant for rural electrification with solar energy in developing countries. In contrast to laboratory and commercial rivalries over which company produces the most thermodynamically efficient solar cells, the firms that manufacture a-Si PV modules for markets in developing countries have focused on lower efficiency but significantly less expensive products (Green

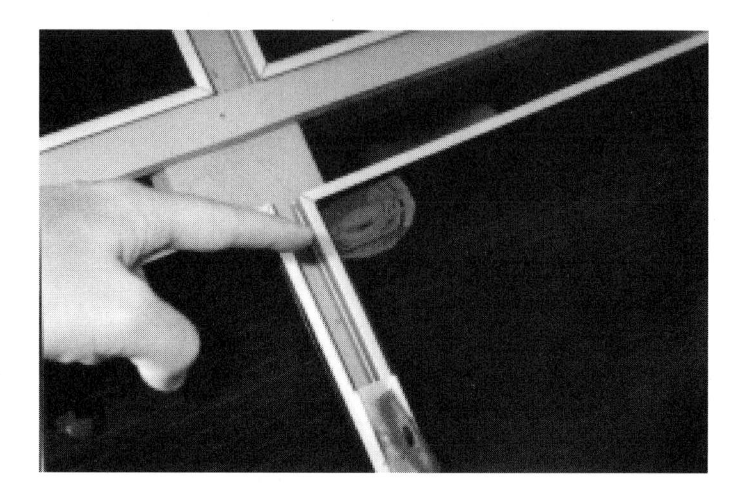

FIGURE 4 Water-intrusion-related delamination in a Brand D a-Si PV module. The actual power output of this 14W rated module was less than 1W.

et al., 2005). The resulting 12 to 20W a-Si PV modules now available in Kenya and elsewhere cost 50 percent less than comparable crystalline silicon (c-Si) PV modules, and are, by far, the best-selling solar products in the region.

The dissemination of a-Si PV technology in Kenya has not, however, been without complications. In an extensive market survey (Figure 5), we found that, although most manufacturers produce high-quality products, one prominent brand performed well below its advertised levels. A previous study in 1999 showed a similar pattern, although for a different brand. Thus, the successful deployment of new technology requires market institutions that ensure quality and protect the public interest. The combination of technical studies of solar equipment performance and analyses of Kenyan market development, socio-cultural dynamics, and regulatory policy has led to progress toward eliminating low-performing products from the market, as well as insights into institutional aspects of renewable energy market development (Acker and Kammen, 1996; Duke et al., 2002; Jacobson, 2004; Kammen, 1995).

MAKING SUSTAINABLE SCIENCE THE NORM

The first step in making sustainable science the norm is to demonstrate that, once funding and a research/action team have been assembled, these projects are no more difficult than traditional research projects. To be effective, however, projects must be neither exclusively in the academic or laboratory setting, nor entirely in the sphere of nonprofit organizations or local governments. To take

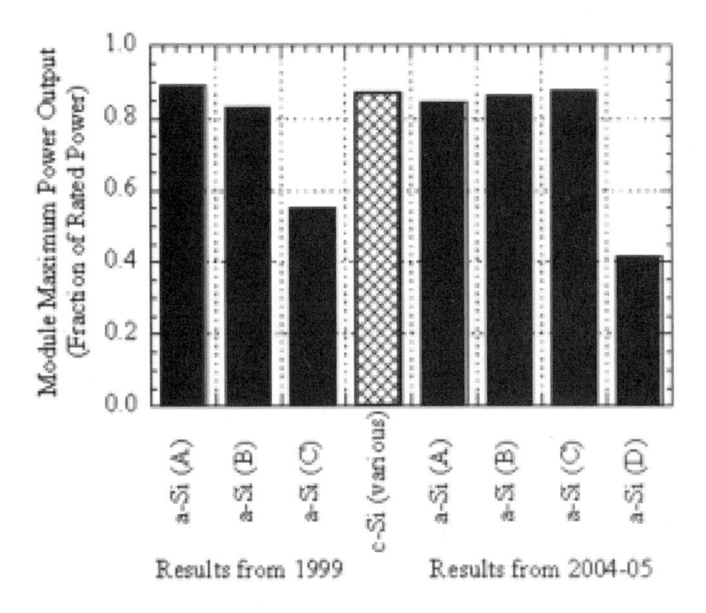

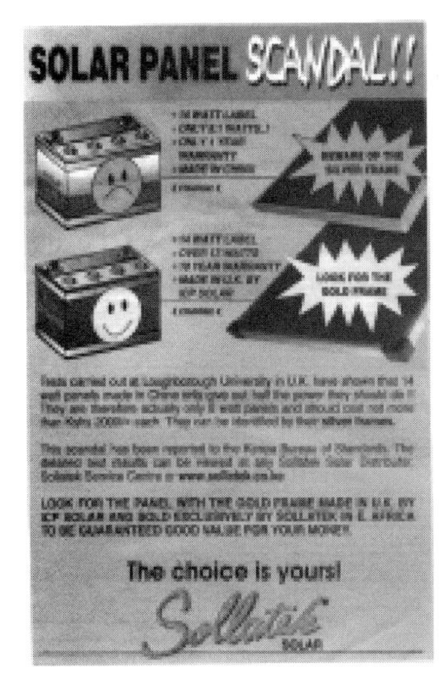

FIGURE 5 Average stabilized maximum power output results from 1999 and 2004–2005 for a-Si solar modules sold in Kenya. Aggregate test results for several brands of c-Si modules are included for comparison. Note that, although most a-Si brands have power output levels similar to the more expensive c-Si modules, some brands perform well below their advertised power ratings. The 1999 test results are based on field measurements of 130 a-Si modules and 17 c-Si modules. The 2004–2005 results involved 20 a-Si modules randomly selected from Kenyan retail shops. The presence of low-performing brands has led to considerable acrimony in the Kenyan solar industry, as indicated in the "Solar Scandal" advertisement from a local newspaper. Following the release in Kenya of the 2004–2005 results, the market presence of Brand D dropped. Source: Jacobson and Kammen, 2005.

maximum advantage of both the emerging science and the implementation capacity for sustainability, we must demonstrate support in each of the disciplines involved, both through actions and funding priorities.

Second, we must make sustainability science a basic precept of teaching in secondary schools, colleges, and postgraduate studies. Pre-college students have already demonstrated a tremendous aptitude for working in interdisciplinary areas. We must nurture and reward this interest with courses in junior high schools, high schools, and colleges on energy, the environment, and the social drivers of resource degradation. In the United States, the Upward Bound Math-Science Program (DOEd, 2005) and Summer Science Program (2005) are models that could be adapted to the theme of sustainability science.

The launch of Sputnik in 1957 initiated an unprecedented mobilization of U.S. science and technology, a lesson in the power of a use-inspired drive to innovate. The Yale Environment Survey found overwhelming interest in energy and environmental sustainability (Yale University, 2005). Contrast that interest with the results of the 3rd International Mathematics and Science Study (TIMSS), in which American secondary school students ranked 19th out of 21 countries in both math and science (NRC, 1997). The TIMMS authors concluded that science and mathematics education in the United States lacked direction, vision, and motivation. Sustainability science could give science, mathematics, and engineering education renewed meaning and immediacy, with paradigm-changing possibilities in both developed and developing nations.

Third, we could establish sustainability awards—modeled after the Ashoka Innovators Awards (2005), the Ansari X Prize (X Prize Foundation, 2005) for the launch of a space vehicle, and the Ashden Awards (2005) for sustainable energy. These awards would bring together partners from developed and developing nations in academia, industry, civil society, and government and would encourage groups to take action on critical sustainability projects. Ideally, sustainability awards, jointly sponsored by private foundations and state or federal governments, would take advantage of the diversity of perspectives and skills that interdisciplinary, international teams would bring together.

Finally, we must address the principal weakness in the economies of many poor nations—a lack of capacity to compete in the global marketplace. Debt forgiveness for impoverished countries in Africa and elsewhere is laudable (Sachs, 2005), but it has already been criticized by African leaders who have noted that aid alone is not a panacea. Estimates of the percentage of overall economic growth from innovation in science and technology, virtually all in industrialized nations, are as high as 90 percent (Solow, 2000). Developing economies would be energized by dramatically increased investment in indigenous innovation. A natural way to do that would be to reward investments in science and technology capacity for sustainable development with additional debt relief or more favorable trade arrangements. This is a perfect time for the

G8 to adopt this plan and assist all nations to invest in environmentally conscious innovation.

ACKNOWLEDGEMENTS

This work was supported by the Energy Foundation; the Class of 1935 of the University of California, Berkeley; the Schatz Energy Research Center; and a grant from the Gordon and Betty Moore Foundation. We thank Robert Bailis and Zia Mian for their insightful comments.

REFERENCES

Acker, R., and D.M. Kammen. 1996. The quiet (energy) revolution: the diffusion of photovoltaic power systems in Kenya. Energy Policy 24(1): 81–111.

Ashden Awards. 2005. The Ashden Awards for Sustainability. Available online at: *http://www.ashdenawards.org.*

Ashoka. 2005. Ashoka Innovators Awards. Available online at: *http://www.ashoka.org.*

Bailis, R., M. Ezzati, and D.M. Kammen. 2005. Mortality and greenhouse gas impacts of biomass and petroleum energy futures in Africa. Science 308(5718): 98–103.

Bush, V. 1945. Science—The Endless Frontier: A Report to the President. Washington, D.C.: U.S. Government Printing Office.

Clarke, T. 2002. Wanted: scientists for sustainability. Nature 418(6900): 812–814.

DOEd (U.S. Department of Education). 2005. Upward Bound Math-Science Program. Available online at: *http://www.ed.gov/programs/triomathsci/index.html.*

Duke, R., A. Jacobson, and D.M. Kammen. 2002. Photovoltaic module quality in the Kenyan solar home systems market. Energy Policy 30(6): 477–499.

ESDA (Energy for Sustainable Development, Africa). 2003. Study on PV Market Chains in East Africa. Report for the World Bank. Nairobi, Kenya: ESDA.

Ezzati, M., and D.M. Kammen. 2001. Indoor air pollution from biomass combustion as a risk factor for acute respiratory infections in Kenya: an exposure-response study. Lancet 358(9282): 619–624.

Ezzati, M., and D.M. Kammen. 2002. Evaluating the health benefits of transitions in household energy technologies in Kenya. Energy Policy 30(10): 815–826.

FSTS (Forum on Science and Technology for Sustainability). 2005. Forum on Science and Technology for Sustainability. Available online at: *http://sustsci.harvard.edu/.*

Green, M.A., K. Emery, D.L. King, S. Igari, and W. Warta. 2005. Short communication: solar cell efficiency tables (version 26). Progress in Photovoltaics: Research and Applications 13(5): 387–392.

Hankins, M. 2000. Energy Services for the World's Poor. Washington, D.C.: ESMAP, World Bank.

Hankins, M., and M. Bess. 1994. Photovoltaic Power to the People: The Kenya Case. Report for the Joint UNDP/World Bank Energy Sector Management Assistance Programme (ESMAP). Washington, D.C.: ESMAP, World Bank.

Hansen, J., L. Nazarenko, R. Ruedy, M. Sato, J. Willis, A. Del Genio, D. Koch, A. Lacis, K. Lo, S. Menon, T. Novakov, J. Perlwitz, G. Russell, G. Schmidt, and N. Tausnev. 2005. Earth's energy imbalance: confirmation and implications. Science 308(5727): 1431–1435.

Jacobson, A. 2004. Connective Power: Solar Electrification and Social Change in Kenya. Unpublished Ph.D. dissertation, University of California, Berkeley.

Jacobson, A., and D.M. Kammen. 2005. Engineering, institutions, and the public interest: evaluating product quality in the Kenyan solar photovoltaics industry. Submitted to Energy Policy.

Jacobson, A., R. Duke, D.M. Kammen, and M. Hankins. 2000. Field Performance Measurements of Amorphous Silicon Photovoltaic Modules in Kenya. In ASES Annual and National Passive Solar Conference Proceedings Combined 2000. Madison, Wis.: American Solar Energy Society.

Kammen, D.M. 1995. Cookstoves for the developing world. Scientific American 273(1): 72–75.

Kammen, D.M., and M.R. Dove. 1997. The virtues of mundane science. Environment 39(6): 10–15, 38–41.

Kates, R.W., W.C. Clark, R. Corell, J.M. Hall, C.C. Jaeger, I. Lowe, J.J. McCarthy, H.J. Schellnhuber, B. Bolin, N.M. Dickson, S. Faucheux, G.C. Gallopin, A. Grübler, B. Huntley, J. Jäger, N.S. Jodha, R.E. Kasperson, A. Mabogunje, P. Matson, H. Mooney, B. Moore III, T. O'Riordan, and U. Svedin. 2001. Sustainability science. Science 292(5517): 641–642.

Kennedy, D. 2003. Editorial: Sustainability and the Commons. Science 302(5652): 1861.

Kennedy, D. 2005. Editorial: The Fight of the Decade. Science 308(5729): 1713.

McMichael, A.J., C.D. Butler, and C. Folke. 2003. New visions for addressing sustainability. Science 302(5652): 1919–1920.

NAS (National Academy of Sciences). 2003. Special section of the July 8 issue devoted to science in support of sustainability. Proceedings of the National Academy of Sciences 100(14).

NRC (National Research Council). 1997. A Splintered Vision: An Investigation of U.S. Science and Mathematics Education, edited by W.H. Schmidt, S.A. Raizen, and C.C. McKnight. Third International Mathematics and Science Study. Washington, D.C.: National Academy Press.

Rich, D.Q., J. Schwartz, M.A. Mittleman, M. Link, H. Luttmann-Gibson, P.J. Catalano, F.E. Speizer, and D.W. Dockery. 2005. Association of short-term ambient air pollution concentrations and ventricular arrhythmias. American Journal of Epidemiology 161(12): 1123–1132.

Sachs, J.D. 2005. Four easy pieces. New York Times, p. A15, June 25, 2005.

SAIP (South African Institute of Physics). 2005. World Conference on Physics and Sustainable Development, Durban, South Africa, 31 October to 2 November 2005. Available online at: *http://www.saip.org.za/physics2005/WCPSD2005.html.*

Schumacher, E.F. 1973. Small Is Beautiful. New York: Harper and Row.

Smith, K.R., S. Mehta, and M. Maeusezahl-Feuz. 2004. Indoor Air Pollution from Household Use of Solid Fuels. Pp. 1435–1493 in Comparative Quantification of Health Risks: Global and Regional Burden of Disease Attributable to Selected Major Risk Factors, edited by M. Ezzati, A.D. Lopez, A. Rodgers, and C.J.L. Murray. Geneva: World Health Organization.

Solow, W.M. 2000. Growth Theory: An Exposition. Oxford, U.K.: Oxford University Press.

Staebler, D.L., and C.R. Wronski. 1977. Reversible conductivity changes in discharge-produced amorphous Si. Applied Physics Letters 31(4): 292–294.

Stokes, D. 1997. Pasteur's Quadrant. Washington, D.C.: Brookings Institute Press.

Su, T., P.C. Taylor, G. Ganguly, and D.E. Carlson. 2002. Direct role of hydrogen in the Staebler-Wronski effect in hydrogenated amorphous silicon. Physical Review Letters 89(1): 015502.

Summer Science Program. 2005. The Summer Science Program. Available online at: *http://www.summerscience.org/home/index.php.*

Swart, R., P. Raskin, J. Robinson, R. Kates, and W.C. Clark. 2002. Critical challenges for sustainability science. Science 297(5589): 1994–1995.

X Prize Foundation. 2005. X Prize. Available online at: *http://www.xprizefoundation.com.*

Yale University. 2005. Yale Center for Environmental Law and Policy. Available online at: *http://www.yale.edu/envirocenter/environmentalpoll.htm.*

Engineering Complex Systems

Introduction

Luis A. Nunes Amaral
Northwestern University
Evanston, Illinois

Kelvin H. Lee
Cornell University
Ithaca, New York

In this brief introduction to complex systems, we distinguish between complicated and complex systems. We also give an overview of the challenges posed by the study of complex systems and a brief description of the collections of concepts and techniques being used to uncover some of their properties

More than the usual session title, this one begs for clarification. For example, when do we *not* study complex systems? Or, at least, very complicated systems? Isn't engineering what engineers do? To clarify what this session is about, it is useful to first understand the differences between complex problems and complicated problems.

Consider the January 2004 announcement by President Bush of a new initiative to send humans to Mars. A lot of heated discussion ensued, but images of a human colony on the Moon, and later on Mars, have made their way into funding agencies, such as DARPA (Defense Advanced Research Projects Agency) and NASA (National Aeronautics and Space Administration), the media, and the collective imagination.

As scientists and engineers, we are trained to check if a problem is likely to have a feasible solution based on present day knowledge before we embark on solving it. The announcement by President Bush prompts at least two important questions. First, will we be able to send humans to Mars? Second, will we be able to build a viable colony on Mars?

Sending humans to Mars poses a number of very *complicated* problems: protection from radiation, including solar wind, during the long flight; the physiologic effects of the absence of gravity; and the landing of a heavy load on a planet with a significant atmosphere and gravitational pull. All of these problems are quite complicated. However, we can work on solving each one separately using, mostly, present day technology and knowledge. In short, we can break the problem into small pieces, solve each piece, and then put the solutions together.

What about building a viable colony on Mars, that is, a colony that can produce its own food, oxygen, energy, and so on? The most natural way to produce food and oxygen is by means of an ecosystem—a stable collection of different species that interact. Building an ecosystem is a complex problem. Its solution cannot be broken into small pieces because an ecosystem only exists in its entirety.

Although it may be difficult to come up with an all-encompassing definition of a complex system, let us attempt it. A complex system is a system with a large number of elements, building blocks, or agents capable of interacting with each other and with their environment. Interactions between elements may occur with immediate neighbors or distant neighbors; agents can be identical or different from each other; they may move in space or occupy fixed positions; and they can be in one of two states or multiple states. The common characteristic of all complex systems is that they display organization without any external organizing principle being applied. The whole is much more that the sum of its parts.

Studies of complex systems have raised some of the most elusive and fascinating questions being investigated by scientists today: how consciousness arises out of the interactions of neurons in the brain and between the brain and its environment; how humans create and learn societal rules; or how DNA orchestrates processes in our cells.

We do not yet have a framework for understanding, designing, or engineering complex systems, such as ecosystems. The difficulty of such undertakings was brought home by the failure of Biosphere 2. Built in the 1980s at a cost of $150 million, Biosphere 2—named in deference to Biosphere 1, the Earth—was originally designed as a closed system environment. From 1991 to 1993, eight "biospherians'" sealed themselves in the glass-enclosed dome and tried to survive on sustainable agriculture and recycling. For reasons still unknown, low crop yields and a 25-percent decrease in oxygen supply quickly led to the failure of the experiment.

The speakers in this session will provide an overview of the theoretical and experimental tools that are enabling us to tackle challenges posed by complex systems in a systematic way. On the experimental side, new high-throughput techniques are revolutionizing our understanding of processes at the cellular level, and large computers are being used to analyze vast amounts of social, economic, and financial data. On the theoretical side, a new tool—network analysis—has been added to a tool kit that already contains nonlinear dynamics, statistical physics, and discrete (agent-based) modeling. Network analysis has already led to significant advances in the modeling and characterization of complex systems.

Talk 1. Complex Networks: Ubiquity, Importance, and Implications, by Alessandro Vespignani

What do metabolic pathways and ecosystems, the Internet, and the propagation of HIV infection have in common? Until a few years ago, the answer would have been very little. In the last few years, a different answer has emerged—they have similar network architectures. Seemingly out of nowhere, in the span of a few years, network theory has become one of the most visible pieces of the body of knowledge that can be applied to the description, analysis, and understanding of complex systems.

Talk 2. Engineering Bacteria for Drug Production, by Jay Keasling

Biological systems are among the most complex systems we know. Even "simple" organisms have evolved elegantly complex chemistry to carry out a variety of functions. Using the tools of modern biology, one can study, create, transfer, and manipulate parts of the chemistry of life. An important goal of this research is the practical application of these techniques to the creation of organisms with new and useful properties to benefit society.

Talk 3. Population Dynamics of Human Language, by Natalia Komarova

Hallmarks of complex systems are adaptability and emergence. Consider how ants find food sources and how their communication methods efficiently solve the problem of the search for and transport of food. A particularly exciting realm in which emergence is of great importance is language acquisition, a topic to which this author has made important contributions.

Talk 4. Agent-Based Modeling as a Decision Making Tool, by Zoltán Toroczkai

Agent-based models are being used to study diverse systems, from ant colonies to the behavior of traders in financial systems to traffic patterns and urban growth to the spread of epidemics. Traffic modeling is one of the most successful applications so far. Indeed, a number of cities now use traffic models to predict and attempt to control traffic patterns.

Complex Networks: Ubiquity, Importance, and Implications

ALESSANDRO VESPIGNANI
Indiana University
Bloomington, Indiana

Thanks to increasingly powerful computers and the informatics revolution, data sets on several large-scale networks can now be gathered and handled systematically. These data can then be used to reveal the structural and functional properties of these networks. Mapping projects of the World Wide Web (WWW) and the physical Internet offered researchers their first opportunity to study the topology and traffic of large-scale networks. Gradually, other studies followed describing networks of practical interest in social science, infrastructure analysis, and epidemiology. These studies, which involved researchers from many different disciplines, have led to a paradigm shift—the *complexity* of large-scale networks has become the central issue in our understanding, characterization, and modeling (Barabási, 2002; Dorogovstev and Mendes, 2003; Newman, 2003; Pastor-Satorras and Vespignani, 2004).

UBIQUITY

In general terms, a network system can be described as a graph with nodes (vertices) that identify the elements of the system; the connecting links (edges) represent the presence of a relation or interaction among these elements. From this very general vantage point, it is easy to perceive that a wide array of systems can be approached in the framework of network theory.

In the first instance, we can provide a rudimentary taxonomy of real-world networks. The two main classes of networks are infrastructure systems and natural or living systems. These classes can be further divided into subgroups. Networks belonging to the class of natural systems can be divided into biological networks, social-system networks, food webs, and ecosystems, just to mention a few. Biological networks can refer to complicated interactions among the genes, proteins, and molecular processes that regulate life. Social networks concern relations between individuals, such as family relationships, friendships, business relationships, and many others (Castells, 2000; Wasserman and Faust, 1994).

Infrastructure networks can readily be divided into two main subgroups, (1) virtual or cyber networks and (2) physical systems. Virtual or cyber networks exist and operate in the digital world of cyberspace. Physical systems include real-world networks, such as energy, transportation, and health care networks. Of course, this is just a rough classification because there are many interrelations and interdependencies among physical infrastructure networks, as well as between physical and digital networks.

The Internet, for instance, is a kind of hybrid network in which cyber features are mixed with physical features. The Internet is composed of physical objects, such as routers (the main computers that enable us to communicate) and transmission lines (the cables that connect computers). On top of this physical layer is a virtual world of software that defines networks, such as WWW, the e-mail network, and peer-to-peer networks.

These information-transfer media are used by hundreds of millions of people, and, similar to the physical Internet, virtual networks have become enormous and intricate as the result of a self-organized growing process. The dynamics of these virtual networks are the outcome of interactions among the individuals that form various communities within the network. In this sense, they include a mixture of complex socio-technical aspects. Other examples of socio-technical networks in which physical and technological constraints interact with social, demographic, and economic factors, include the worldwide airport and power-distribution networks.

COMPLEXITY

To understand complex networks, we must make a critical distinction between "complex networks" and "complicated networks." The characteristic features and behavior of complex systems differ significantly from those of merely complicated systems. A minimal definition of complexity includes two main features: (1) complex systems exhibit complications and heterogeneities on all scales possible within the physical constraints of the system; (2) these complications and heterogeneities are spontaneous outcomes of interactions among the constituent units of the system; in other words, they are emergent phenomena.

It is easy to see that WWW, the Internet, and the airport network are systems that grow in time by following complicated dynamic rules without a blueprint or global supervision. The same can be said of many social and biological networks. All of them are self-organizing systems that, at the end of their evolution, show an emergent architecture with unexpected properties and regularities.

If complexity is inherent in the emergence of complications on all scales, one might wonder what a signature of complexity would be in real-world networks. The first clue is the high level of heterogeneity in the degree of vertices, that is, the number of connections, *k*, to each vertex. This feature can easily be understood by visualizing airport networks and the "hub" policy that most airlines have adopted. The same arrangement can be perceived in many other networks in which the presence of "hubs" is a natural consequence of different factors, such as popularity, strategies, and optimization. In the WWW, for instance, some pages become so popular they are pointed to by thousands of other pages; in general, however, most documents remain practically unknown.

The presence of hubs and connectivity ordering have a more dramatic manifestation than was initially thought (Albert and Barabási, 2002; Barabási and Albert, 1999). This can be observed by studying the degree distribution $P(k)$, i.e., the probability that any given vertex has *k* connection to other vertices. For a large number of networks this distribution is very different from the one usually encountered in Poissonian random graphs and exhibits heavy-tail often approximated by power-law forms $P(k) \sim k^{-\gamma}$. A peculiarity of distributions with heavy tails is that the average behavior of the system is not typical (Figure 1).

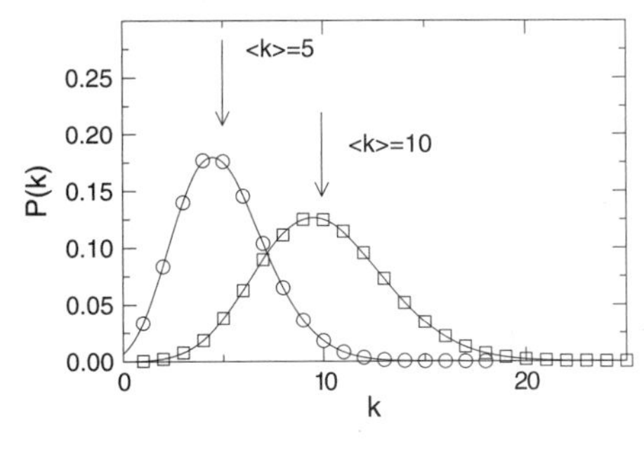

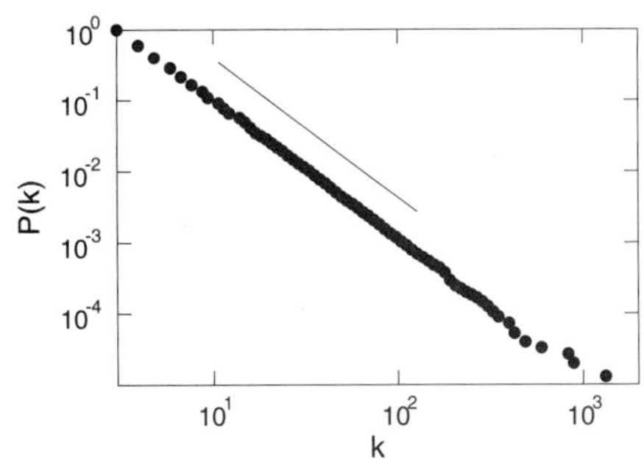

FIGURE 1 Poisson distributions (top) and a power-law distribution typical of scale-free graphs (bottom). Poisson distributions are peaked around the average value <k> and decay exponentially at large degree values. The power-law distribution $P(k) \sim k^{-\gamma}$ acquires a linear behavior on the double logarithmic scale. The slope identifies the value of the γ exponent. It is worth stressing the broad spectrum of the power-law distribution that extends up to very high values in k.

In mathematical terms, the heavy-tail property translates into a very high level of fluctuation, with the divergence of the standard deviation of the degree distribution limited only by the finite size of the system (Albert and Barabási, 2002; Amaral et al., 2000; Barabási and Albert, 1999). We are thus in the presence of structures with fluctuations and complications that extend over all possible scales allowed by the physical size of the system. In short, these are complex systems.

IMPORTANCE

Heavy tails and heterogeneity appear to be common characteristics of a large number of real-world networks, along with other complex topological features, such as the presence of communities, motifs, hierarchies, and modular ordering. The evidence that a complex topology is shared by many complex evolving networks cannot be considered incidental. Rather, it suggests a general principle that might explain the emergence of this architecture in very different contexts. From this perspective, it would be useful to have a theoretical understanding that might reveal the general principles that underlie the formation of a network.

When looking at networks from the point of view of complex systems, the focus is on the microscopic processes that govern the appearance and disappearance of vertices and links. Thus, attempts to model and understand the origin of observed topological properties of real-world networks require a radical change in perspective to predictions of the large-scale properties and behavior of the system on the basis of dynamic interactions among the constituent units. For this reason, recent activity has been focused on the development of dynamic models for networks. This intense activity has generated a vast field of research whose results and advances are providing new techniques for approaching conceptual and practical problems in the field of networked systems (Amaral et al., 2000; Barabási, 2002; Barabási and Albert, 1999; Dorogovstev and Mendes, 2003; Newman, 2003; Pastor-Satorras and Vespignani, 2004).

IMPLICATIONS

Advances in the understanding of large complex networks have also generated a great deal of interest in the potential implications of complex properties for the engineering, optimization, and protection of networks. These problems are emerging as fundamental issues that are relevant beyond the usual basic-science perspective. For instance, the complexity of networks has important consequences for the empirical analysis of the robustness of a network to failure or attack.

A natural question to ask in this context concerns the maximum amount of damage a network can withstand (i.e., the threshold value of the removal density

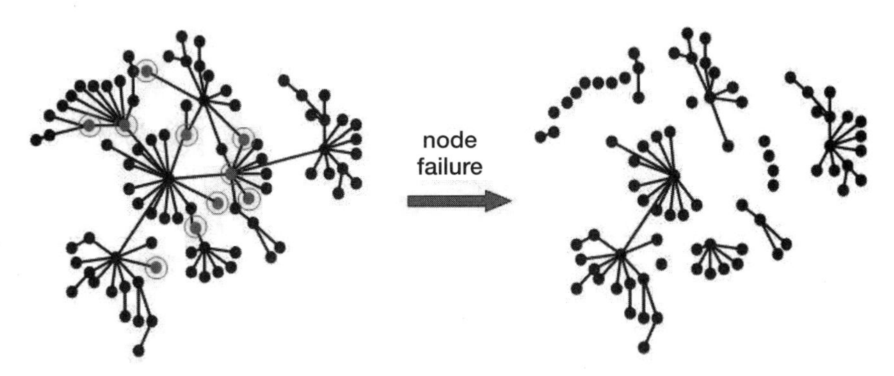

FIGURE 2 Analysis of topological resilience of a network to attacks and failures. If the circled nodes in the network (left) are removed as a result of malfunction or attack, the network is broken up into several fragmented components.

of vertices above which a network can be considered compromised). Contrary to non-complex networks, heavy-tailed networks have two levels of robustness in the face of component failures. First, they are extremely robust to the loss of a large number of randomly selected vertices. Second, they are extremely fragile in response to a targeted attack (Figure 2) (Albert et al., 2000; Cohen et al., 2000).

Studies of complex networks have also revealed important insights on the properties of disease-spreading in highly heterogeneous networks (Lijeros et al., 2001; Schneeberger et al., 2004). Indeed, the presence of heavy-tailed connectivity patterns changes the epidemic framework dramatically. In less complex networks, it is possible to show, on a general basis, that if the rate of spread during an epidemic—roughly speaking the disease transmission probability—is below a given threshold value, the epidemic will die out in a very short time. However, in scale-free networks, no matter the spreading rate, there is a finite probability that the infection will cause a major outbreak that pervades the system or will reach a long-lasting steady state. Thus, heavy-tailed networks lack epidemic thresholds (Pastor-Satorras and Vespignani, 2001a). Interestingly, the absence of an epidemic threshold corresponds to a general inadequacy of uniform immunization policies. Nevertheless, it is possible to take advantage of the connectivity pattern of heavy-tailed networks. One can show in mathematical terms that the protection of just a tiny fraction of the most connected individuals dramatically increases the tolerance level of the whole population to the disease (Cohen et al., 2003; Dezso and Barabási, 2001; Pastor-Satorras and Vespignani, 2001b).

Finally, complexity features also affect the dynamics of information or traffic flow on the network structure. The resilience, or robustness, of networks is a dynamic process that is affected by the time response of elements to different damage configurations. For instance, when a router or connection fails, the Internet responds very quickly by updating the routing tables of other routers in the neighborhood of the failure point. In most cases, this adaptive response is able to circumscribe the damage, but in some cases failures may cascade through the network, causing far more disruption than one would expect from the initial cause (Lee et al., 2005; Moreno et al., 2003; Motter and Lai, 2002).

Cascading failure is typical of complex systems, in which emergent properties imply events and information flow over a wide range of length and time scales. In other words, small perturbations have a finite probability of triggering a system-wide response, so-called critical behavior. This happens through chains of events that eventually involve a large macroscopic part of the system and, in some cases, lead to global failure.

It is important to realize that in large networked systems this property is inherent in the system complexity and cannot be eliminated by local reinforcement or technological updates. We can vary the proportion of small and large events, but we must learn to live with appreciable probabilities of very large events. We must deal with the inherent complexity of the real world.

REFERENCES

Albert, R., and A.-L. Barabási. 2002. Statistical mechanics of complex networks. Reviews of Modern Physics 74(1): 47–97.

Albert, R., H. Jeong, and A.-L. Barabási. 2000. Error and attack tolerance of complex networks. Nature 406 (6794): 378–382.

Amaral, L.A.N., A. Scala, M. Barthélémy, and H.E. Stanley. 2000. Classes of small world networks. Proceedings of the National Academy of Sciences 97(21): 11149–11152.

Barabási, A.-L. 2002. Linked. Jackson, Tenn.: Perseus Books Group.

Barabási, A.-L., and R. Albert. 1999. Emergence of scaling in random networks. Science 286(5439): 509–512.

Castells, M. 2000. The Rise of the Network Society, vols. 1 and 2. Oxford, U.K.: Blackwell.

Cohen, R., K. Erez, D. ben-Avraham, and S. Havlin. 2000. Resilience of the Internet to random breakdowns. Physical Review Letters 85 (21): 4626–4629.

Cohen, R., S. Havlin, and D. ben-Avraham. 2003. Efficient immunization strategies for computer networks and populations. Physical Review Letters 91(24): 247201.

Dezso, Z., and A.-L. Barabási. 2001. Halting viruses in scale-free networks. Physical Review E 65(5): 055103.

Dorogovstev, S.N., and J.F.F. Mendes. 2003. Evolution of Networks. Oxford, U.K.: Oxford University Press.

Lee, E.J., K.-I. Goh, B. Kahng, and D. Kim. 2005. Robustness of the avalanche dynamics in data packet transport on scale-free networks. Physics Review E 71(5): 056108.

Lijeros, F., C.R. Edling, L.A.N. Amaral, H.E. Stanley, and Y. Åberg. 2001. The web of human sexual contacts. Nature 411(6840): 907–908.

Moreno, Y., R. Pastor-Satorras, A. Vázquez, and A. Vespignani. 2003. Critical load and congestion instabilities in scale-free networks. Europhysics Letters 62(2): 292–298.

Motter, A.E., and Y.C. Lai. 2002. Cascade-based attacks on complex networks. Physics Review E 66(6): 065102.

Newman, M.E.J. 2003. Structure and function of complex networks. SIAM Review 45(2): 167–256.

Pastor-Satorras, R., and A.Vespignani. 2001a. Epidemic spreading in scale-free networks. Physical Review Letters 86(14): 3200–3203.

Pastor-Satorras, R., and A. Vespignani. 2001b. Immunization of complex networks. Physical Review E 65(3): 036104.

Pastor-Satorras, R., and A. Vespignani. 2004. Evolution and Structure of the Internet. Cambridge, U.K.: Cambridge University Press.

Schneeberger. A., C.H. Mercer, S.A.J. Gregson, N.M. Ferguson, C.A. Nyamukapa, R.M. Anderson, A.M. Johnson, and G.P. Garnett. 2004. Scale-free networks and sexually transmitted diseases. Sexually Transmitted Diseases 31(6): 380–387.

Wasserman, S., and K. Faust. 1994. Social Network Analysis: Methods and Applications. Cambridge, U.K.: Cambridge University Press.

The Promise of Synthetic Biology

JAY KEASLING
University of California, Berkeley

It has been estimated that for every successful drug compound, 5,000 to 10,000 compounds must be introduced into the drug-discovery pipeline. On average, it takes $802 million and 10 to 15 years to develop a successful drug. Given this very low success rate and the incredibly high costs, drug companies must introduce as many drug candidates into their pipelines as possible.

Natural products have been important sources of drug leads; as much as 60 percent of successful drugs are of natural origin (Cragg et al., 1997), and some of the most potent natural products have been used as anticancer, antibacterial, and antifungal drugs. However, most natural products evolved for purposes other than the treatment of human disease. Thus, even though they can sometimes function as human therapeutics, their pharmacological properties may not be optimal. Furthermore, many are produced in miniscule amounts in their native hosts, thus making them expensive to harvest.

Organic chemistry methodologies are widely used to synthesize pharmaceuticals (of natural origin or not) and functionalize pharmaceutically relevant natural products. With appropriate protection and deprotection steps, chiral centers and functionalities can be introduced into molecules with precision. With the advent of combinatorial chemical synthesis, researchers have been able to construct entire families of molecules substituted at several positions with several different substituents, thus allowing drug companies to fill drug-discovery pipelines with variations of promising leads.

Despite the creation of complicated molecules made possible by advances in organic synthesis methodologies, the performance of these molecules is hardly comparable to the ease, specificity, and "green-ness" of enzymes. Indeed, many organic synthesis routes now incorporate one or more enzymes to perform transformations that are particularly difficult using nonenzymatic routes. Furthermore, enzymes are now being used for the in vitro, combinatorial functionalization of complex molecules. The next logical step in the synthesis of chemotherapeutics is the use of enzymes for combinatorial synthesis inside the cell, which would allow the production of drug candidates from inexpensive starting materials and avoid the need for purification of enzymes, which may be necessary for in vitro synthesis.

BIOLOGICAL ENGINEERING FOR THE SYNTHESIS OF DRUGS

Rich, versatile biological systems are ideally suited to solving some of the world's most significant challenges, such as converting cheap, renewable resources into energy-rich molecules; producing high-quality, inexpensive drugs to fight disease; detecting and destroying chemical or biological agents; and remediating polluted sites. Over the years, significant strides have been made in engineering microorganisms to produce ethanol, bulk chemicals, and valuable drugs from inexpensive starting materials; to detect and degrade nerve agents as well as less toxic organic pollutants; and to accumulate metals and reduce radionuclides.

However, meeting these biological engineering challenges requires long development times, largely because of a lack of useful tools that would enable engineers to easily and predictably reprogram existing systems, let alone build new enzymes, signal transduction pathways, genetic circuits, and, eventually, whole cells. The ready availability of these tools would drastically alter the biotechnology industry, leading to less expensive pharmaceuticals, renewable energy, and biological solutions to problems that do not currently offer sufficient monetary returns to justify the high cost of biological research.

Most of the biological engineering tools currently available to scientists and engineers have not changed significantly since genetic engineering began in the 1970s. Biologists still use natural, gene-expression control systems (promoters with cognate repressors/activators). The ability to place a single heterologous gene under the control of one of these native promoters and produce large quantities of a protein of interest is the basis for the modern biotechnology industry.

Although redesigned biological control systems have been generally effective for their intended purposes (controlling rather roughly the expression of a single gene or a few genes), not surprisingly they are often inadequate for more complicated engineering tasks (e.g., controlling very large, heterologous, metabolic pathways or signal transduction systems). In addition, these borrowed "biological parts" retain many of the features that were beneficial in their native forms but make them difficult to use for purposes other than the ones for which they evolved. Well characterized standard biological parts, and larger devices made from such parts, would make biological engineering more predictable and enable the construction and integration of larger systems than is currently possible.

In almost every other field of engineering, standards have been developed for building large integrated systems by assembling components from various manufacturers. However, biologists and engineers have not yet defined standards for the parts that might allow them to build larger biological devices. The design and construction of new devices (e.g., genetic-control systems) would benefit greatly from standards governing how various parts (e.g., regulatory proteins, promoters, ribosome binding sites) should interact and be assembled. Setting standards would also encourage manufacturing firms to develop parts.

Biological engineering has been held back because many of the most effective biological parts (promoters, genes, plasmids, etc.) have been patented and are available only to companies that can afford the royalty payments. This has not only increased the cost of drug development, but also hampered the development of new biological solutions to problems that may not have significant monetary payoffs (basically, anything other than drug development). Open-source biological parts, devices, and eventually whole cells would reduce the cost of engineering biological systems, make biological engineering more predictable, and encourage the development of novel biological solutions to some of our most challenging problems. The development of open-source biological technology would improve awareness of, and minimize possible future biological risks, in the same way that open-source software tends to promote a constructive and responsive community of users and developers.

SYNTHETIC BIOLOGY

Synthetic biology is the design and construction of new biological entities, such as enzymes, genetic circuits, and cells, or the redesign of existing biological systems. The goal of synthetic biology, which builds on advances in molecular, cellular, and systems biology, is to transform biology in the same way that synthesis transformed chemistry and integrated circuit design transformed computing. The element that distinguishes synthetic biology from traditional molecular and cellular biology is the focus on (1) the design and construction of core components (parts of enzymes, genetic circuits, metabolic pathways, etc.) that can be modeled, understood, and engineered to meet specific performance criteria, and (2) the assembly of these smaller parts and devices into larger integrated systems to solve specific problems. Just as engineers now design integrated circuits based on the known physical properties of materials and then fabricate functioning circuits and entire processors (with relatively high reliability), synthetic biologists will soon design and build engineered biological systems.

Unlike many other areas of engineering, however, biology is nonlinear and less predictable, and much less is known about parts and how they interact. Hence, the overwhelming physical details of natural biology (gene sequences, protein properties, biological systems) must be organized and recast via a set of design rules that hide information and manage complexity, thereby enabling the engineering of multicomponent integrated biological systems. Only when this is accomplished will designs of significant scale be possible.

Synthetic biology arose from four different intellectual premises. The first is the scientific idea that a practical test of understanding is the ability to reconstitute a functional system from its basic parts. Using synthetic biology, scientists are testing models of how biology works by building systems based on models and measuring differences between expectations and observations. Second, some consider biology an extension of chemistry, and thus synthetic biology can be considered an extension of synthetic chemistry. Attempts to manipulate living systems at the molecular level will likely lead to a better understanding, and new types, of biological components and systems. Third, natural living systems evolved to ensure their continued existence; they are not optimized for human understanding and intention. By thoughtfully redesigning natural living systems, it is possible simultaneously to test our current understanding and potentially implement engineered systems that are easier to interact with and study. Fourth, biology can be used as a technology, and biotechnology, broadly redefined, includes the engineering of integrated biological systems for the purposes of processing information, producing energy, manufacturing chemicals, and fabricating materials.

Although the emergence of the discipline of synthetic biology was motivated by these agendas, progress has only been practical since the recent advent of two foundational technologies, DNA sequencing, which has increased our understanding of the components and organization of natural biological systems, and synthesis, which has enabled us to begin to test the designs of (1) new, synthetic biological parts (Allert et al., 2004; Basu et al., 2004; Becskei and Serrano, 2000; Cane et al., 2002; Datsenko and Wanner, 2000; De Luca and Laflamme, 2001; Dwyer and Hellinga, 2004; Gardner and Collins, 2000; Gardner et al., 2000; Geerlings et al., 2001; Gerasimenko et al., 2002; Godfrin-Estevenon et al., 2002; Guet et al., 2002; Kobayashi et al., 2004; McDaniel et al., 1997) and (2) new biological systems (Bignell and Thomas, 2001; Blake and Isaacs, 2004; Hughes and Shanks, 2002; Iijima et al., 2004; Irmler et al., 2000; Judd et al., 2000; Kumar et al., 2004; Le Borgne et al., 2001; Martin et al., 2001, 2002, 2003; Okamoto et al., 2004). Each of these examples demonstrates the incredible potential of synthetic biology, as well as the foundational scientific and engineering challenges that must be met for the engineering of biology to become routine.

REFERENCES

Allert, M., S.S. Rizk, L.L. Looger, and H.W. Hellinga. 2004. Computational design of receptors for an organophosphate surrogate of the nerve agent soman. Proceedings of the National Academy of Sciences 101(21): 7907–7912.

Basu, S., R. Mehreja, S. Thiberge, M.-T. Chen, and R. Weiss. 2004. Spatiotemporal control of gene expression with pulse-generating networks. Proceedings of the National Academy of Sciences 101(17): 6355–6360.

Becskei, A., and L. Serrano. 2000. Engineering stability in gene networks by autoregulation. Nature 405(6786): 590–593.

Bignell, C., and C.M. Thomas. 2001. The bacterial ParA-ParB partitioning proteins. Journal of Biotechnology 91(1): 1–34.

Blake, W.J., and F.J. Isaacs. 2004. Synthetic biology evolves. Trends in Biotechnology 22(7): 321–324.

Cane, D.E., F. Kudo, K. Kinoshita, and C. Khosla. 2002. Precursor-directed biosynthesis: biochemical basis of the remarkable selectivity of the erythromycin polyketide synthase toward unsaturated triketides. Chemistry and Biology 9(1): 131–142.

Cragg, G.M., D.J. Newman, and K.M. Snader. 1997. Natural products in drug discovery and development. Journal of Natural Products 60(1): 52–60.

Datsenko, K.A., and B.L. Wanner. 2000. One-step inactivation of chromosomal genes in *Escherichia coli* K-12 using PCR products. Proceedings of the National Academy of Sciences 97(12): 6640–6645.

De Luca, V., and P. Laflamme. 2001. The expanding universe of alkaloid biosynthesis. Current Opinion in Plant Biology 4(3): 225–233.

Dwyer, M.A., and H.W. Hellinga. 2004. Periplasmic binding proteins: a versatile superfamily for protein engineering. Current Opinion in Structural Biology 14(4): 495–504.

Gardner, T.S., and J.J. Collins. 2000. Neutralizing noise in gene networks. Nature 405(6786): 520–521.

Gardner, T.S., C.R. Cantor, and J.J. Collins. 2000. Construction of a genetic toggle switch in *Escherichia coli*. Nature 403(6767): 339–342.

Geerlings, A., F.J. Redondo, A. Contin, J. Memelink, R. van der Heijden, and R. Verpoorte. 2001. Biotransformation of tryptamine and secologanin into plant terpenoid indole alkaloids by transgenic yeast. Applied Microbiology Biotechnology 56(3-4): 420–424.

Gerasimenko, I., Y. Sheludko, X. Ma, and J. Stockigt. 2002. Heterologous expression of a Rauvolfia cDNA encoding strictosidine glucosidase, a biosynthetic key to over 2000 monoterpenoid indole alkaloids. European Journal of Biochemistry 269(8): 2204–2213.

Godfrin-Estevenon, A.M., F. Pasta, and D. Lane. 2002. The parAB gene products of *Pseudomonas putida* exhibit partition activity in both *P. putida* and *Escherichia coli*. Molecular Microbiology 43(1): 39–49.

Guet, C.C., M.B. Elowitz, W. Hsing, and S. Leibler. 2002. Combinatorial synthesis of genetic networks. Science 296(5572): 1466–1470.

Hughes, E.H., and J.V. Shanks. 2002. Metabolic engineering of plants for alkaloid production. Metabolic Engineering 4(1): 41–48.

Iijima, Y., D.R. Gang, E. Fridman, E. Lewinsohn, and E. Pichersky. 2004. Characterization of geraniol synthase from the peltate glands of sweet basil. Plant Physiology 134(1): 370–379.

Irmler, S., G. Schroder, B. St. Pierre, N.P. Crouch, M. Hotze, J. Schmidt, D. Strack, U. Matern, and J. Schroder. 2000. Indole alkaloid biosynthesis in Catharanthus roseus: new enzyme activities and identification of cytochrome P450 CYP72A1 as secologanin synthase. The Plant Journal 24(6): 797–804.

Judd, E.M., M.T. Laub, and H.H. McAdams. 2000. Toggles and oscillators: new genetic circuit designs. Bioessays 22(6): 507–509.

Kobayashi, H., M. Kaern, M. Araki, K. Chung, T.S. Gardner, C.R. Cantor, and J.J. Collins. 2004. Programmable cells: interfacing natural and engineered gene networks. Proceedings of the National Academy of Sciences 101(22): 8414–8419.

Kumar, P., C. Khosla, and Y. Tang. 2004. Manipulation and analysis of polyketide synthases. Methods in Enzymology 388: 269–293.

Le Borgne, S., B. Palmeros, F. Bolivar, and G. Gosset. 2001. Improvement of the pBRINT-Ts plasmid family to obtain marker-free chromosomal insertion of cloned DNA in E. coli. Biotechniques 30(2): 252–254, 256.

Martin, V.J.J., D.J. Pitera, S.T. Withers, J.D. Newman, and J.D. Keasling. 2003. Engineering the mevalonate pathway in Escherichia coli for production of terpenoids. Nature Biotechnology 21(7): 796–802.

Martin, V.J.J., C.D. Smolke, and J.D. Keasling. 2002. Redesigning cells for production of complex organic molecules. ASM News 68: 336–343.

Martin, V.J.J., Y. Yoshikuni, and J.D. Keasling. 2001. The in vivo synthesis of plant sesquiterpenes in Escherichia coli. Biotechnology and Bioengineering 75(5): 497–503.

McDaniel, R., C.M. Kao, S.J. Hwang, and C. Khosla. 1997. Engineered intermodular and intramodular polyketide synthase fusions. Chemistry and Biology 4(9): 667–674.

Okamoto, A., K. Tanaka, and I. Saito. 2004. DNA logic gates. Journal of the American Chemical Society 126(30): 9458–9463.

Population Dynamics of Human Language: A Complex System

NATALIA L. KOMAROVA
University of California, Irvine

In the course of natural history, evolution has made several great innovative "inventions," such as nucleic acids, proteins, cells, chromosomes, multi-cellular organisms, and the nervous system. The last "invention," which truly revolutionized the very rules of evolution, is language. It gives humans an unprecedented possibility of transmitting information from generation to generation, not by the "traditional" means of a genetic code, but by talking. This new mode of cross-generational information transfer has given rise to so-called "cultural evolution." Through cultural evolution, language is responsible for a big part of our being "human." Language, and cultural evolution, are also shaping history and changing the rules of biology. Without exaggeration, language is one of the most fascinating traits of *Homo sapiens*.

The study of language and grammar dates back to classical India and Greece. In the eighteenth century, the discovery of the Indo-European language family led to the surprising realization that very different languages may be related to each other; this was the beginning of historical linguistics. Formal language theory, which emerged only in the 20th century (Chomsky, 1956, 1957; Harrison, 1978), is an attempt to describe the rules a speaker uses to generate linguistic forms (descriptive adequacy) and to explain how language competence emerges in the human brain (explanatory adequacy). Language theory has been supported by advances in the mathematical and computational analysis of language acquisition, a field that became known as learning theory. Currently, efforts are being focused on bringing linguistic inquiry into contact with various disciplines of biology, including neurobiology (Deacon, 1997; Vargha-Khadem et al., 1998), animal behavior (Dunbar, 1996; Fitch, 2000; Hauser, 1996), evolution (Aitchinson, 1996; Batali, 1994; Bickerton, 1990; Hawkins and Gell-Mann, 1992; Hurford et al., 1998; Jackendoff, 1999; Knight et al., 2000; Lieberman, 1984, 1991; Maynard Smith and Szathmary, 1995; Pinker and Bloom, 1990) and genetics (Gopnik and Crago, 1991; Lai et al., 2001). The goal of these interdisciplinary studies is to approach language as a product of evolution and as the extended phenotype of a species of primates.

In the past decade there has been an explosion of interest in computational aspects of the evolution of language (Cangelosi and Parisi, 2001; Christiansen and Kirby, 2003). A great many efforts, across a wide range of disciplines, are now focused on answering such questions as why language is the way it is and how it got that way. Various approaches to these questions have been suggested, including viewing language as a *complex adaptive system* (Steels, 2000). Levin (2002) identified the following defining properties of a general complex adaptive system:

1. They consist of a number of different components.
2. The components interact with each other with some degree of localization.
3. An autonomous process uses the outcomes of these interactions to select a subset of components for replication and/or enhancement.

Property 3, a signature of "biology" in a complex adaptive system, includes replication (which implies a degree of variability) and Darwinian selection. A mathematical problem posed by such systems is to find the outcome (or, more generally, describe the dynamics) of a competition in which the set of players changes, depending on the current state of affairs. New players come in, and their "strategies" (or properties) are drawn from a huge set of possibilities.

The main idea in this approach to language evolution can be stated as follows. There is a population of individuals (neural networks [Oliphant, 1999; Smith, 2002], agents [Steels, 2001; Steels and Kaplan, 1998], organisms in a foraging environment [Cangelosi, 2001]) who communicate with each other. Each individual is characterized by parameters that define its phenotype. These usually include the individual's ability to speak, vocabulary, ability to learn, and other characteristics important for communication (and sometimes other features like the life span or onset of the reproductive age [Hurford and Kirby, 1998]).

The results of communication between individuals are assessed in some way. Rounds of communication are followed by rounds of "update," which may mimic biological reproduction (an individual is replaced by its offspring), or learning (the individual's vocabulary or grammatical rules are changed/updated). Various numerical techniques are used to model the dynamics of reproducing and learning individuals, such as genetic algorithms. The initial condition usually assumes no common communication system in the population. After a number of rounds of update/replication, the state of communication ability is evaluated again.

Two questions are often addressed. First, under what circumstances does a common communication system arise in a population of interacting individuals? And second, what are the conditions under which such a communication system can be maintained?

WHAT IS UNIVERSAL GRAMMAR AND WHY DO WE NEED IT?

Learning is inductive inference. The learner is presented with data and must infer the rules that generate these data. The difference between "learning" and "memorization" is the ability to generalize beyond one's own experience to novel circumstances. In the context of language, the child learner will generalize to novel sentences never heard before. Any child can produce and understand sentences that are not part of his/her previous linguistic experience.

Children develop grammatical competence spontaneously without formal training. All they need is interaction with people and exposure to normal language use. In other words, a child hears grammatical sentences and then constructs an internal representation of the rules that generate grammatical sentences. Chomsky pointed out that the evidence available to the child does not uniquely determine the underlying grammatical rules (Chomsky, 1965, 1972), a phenomenon called the "poverty of stimulus" (Wexler and Culicover, 1980). The "paradox of language acquisition" is that children nevertheless reliably achieve correct grammatical competence (Jackendoff, 1997, 2001). How is this possible?

The proposed solution of the paradox is that children learn correct grammar by choosing from a restricted set of candidate grammars. The structure of this restricted set is "universal grammar." Mathematical learning theory proves the "necessity" of a universal grammar. Discovering properties of universal grammar and particular human learning algorithms requires the empirical study of neurobiological and cognitive functions of the human brain involved in language acquisition. Some aspects of universal grammar, however, might be revealed by studying common features of existing human languages. This has been a major goal of linguistic research during the last several decades.

In our modeling approach, we use the concept and some properties of universal grammar to formulate the mathematical theory of language evolution. We assume that universal grammar has a rule system that generates a set (or a search space) of grammars, $\{G_1, G_2, \ldots, G_n\}$. These grammars can be constructed by the language learner as potential candidates for the grammar that needs to be learned; the learner cannot end up with a grammar that is not part of this search space. In this sense, universal grammar contains the possibility of learning all human languages (and many more). Figure 1 illustrates this process of language

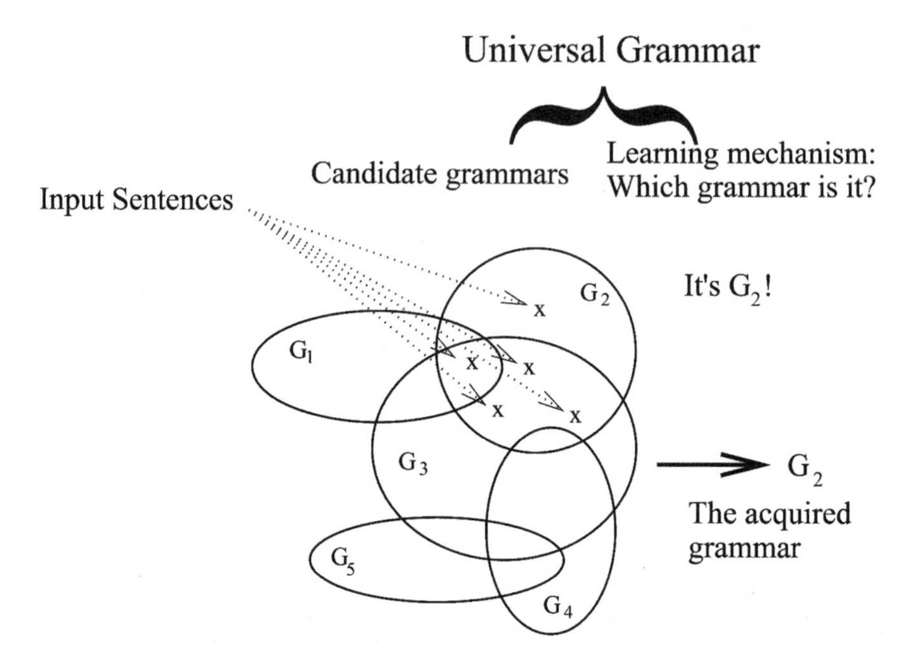

FIGURE 1 Universal grammar specifies the search space of candidate grammars and the learning procedure for evaluating input sentences. The basic idea is that the child has an innate expectation of grammar (for example, a finite number of candidate grammars) and then chooses a particular candidate grammar that is compatible with the input. Source: Komarova and Nowak, 2001c. Reprinted with permission.

acquisition. The learner has a mechanism to evaluate input sentences and to choose one of the candidate grammars in his search space.

A MATHEMATICAL FORMULATION OF LANGUAGE EVOLUTION

Our approach differs from many others in that we use mathematical, *analytical* tools to address questions of language origins and evolution (Komarova and Nowak, 2001a,b, 2003; Komarova et al., 2001; Nowak and Komarova, 2001; Nowak et al., 2001, 2002). We assume that each individual has universal grammar, which allows him/her to learn any language in a (finite but large) set, $\{G_1, \ldots, G_n\}$.

In classical learning theory, which usually focuses on an isolated teacher-learner pair, there is a collection of concepts (grammars), $G_1, \ldots, G_n$, and words (or sample sentences, for learning a grammar) that refer to these concepts, sometimes ambiguously. The teacher generates a stream of words, referring to, say, concept G_2. This is not known to the student, but he must learn by guessing some concept, G_i, and checking for consistency with the teacher's input. A typical question of interest is how quickly a given method converges with the truth. Stated in the terminology of learning languages, the question becomes how many samples, N_δ, a given learning algorithm typically requires to learn the correct language with probability $1-\delta$. Questions of this type for specific learning mechanisms are interesting mathematical problems (e.g., the treatment of the so-called memoryless learner in Komarova and Rivin, 2003).

Next, imagine a population of learners, all equipped with a given learning algorithm. The question now becomes how many samples, N_δ, individual learners in a population need for the fraction $1-\delta$ of the population to converge to a common language. The answer will, of course, depend on the specifics of the population dynamics.

Borrowing from population biology, we can define the fitness of speakers of different grammars. We denote by *sij* the probability that a speaker who uses grammar G_i formulates a sentence that is compatible with grammar G_j. The matrix $\{s_{ij}\}$ describes the pair-wise similarity among the n grammars, $0 \leq s_{ij} \leq 1$. We assume there is a reward for mutual understanding. The payoff for an individual using G_i communicating with an individual using G_j is given by $a_{ij} = (1/2)(s_{ij} + s_{ij})$. This is the average probability that G_i generates a sentence that is parsed by G_j and vice versa. We denote by x_i the frequency of individuals who use grammar G_i; the vector $\mathbf{x} = \{x_1, \ldots, x_n\}$ is defined on the simplex,

$$\sum_{i=1}^{n} x_i = 1.$$

The average payoff of each of these individuals is given by $\mathbf{f} = \hat{A}\mathbf{x}$, where $\hat{A} = \{a_{ij}\}$ is a symmetric matrix. Payoff translates into fitness—individuals with a higher payoff produce more offspring. Note that the fitness of individuals strongly depends on the current composition of the population. Such is the nature of communication.

Another biological concept based on the theory of Darwinian evolution is variability. The “mutation rates” are defined as follows: denote by Q_{ij} the probability that a child learning from a parent with grammar G_i will end up speaking grammar G_j. $\hat{Q} = \{Q_{ij}\}$ is a stochastic matrix (its rows add up to one). Interestingly, the findings related to individual teacher-learner pairs can be incorporated in a natural way into the matrix, $\hat{Q}$.

The last component of the model is the update rule for the evolutionary dynamics. The simplest rule is a deterministic equation, where each variable has a meaning of its ensemble average and the noise is neglected. This can be written by analogy with the well-known quasispecies equation (Eigen and Schuster, 1979), except it has a higher degree of nonlinearity (a consequence of the population-dependent fitness). We have,

$$\dot{x}_i = \sum_{j=1}^{N} x_j f_j Q_{ji} - \phi x_i, 1, \ldots$$

Here ϕ = (**x, f**) is the average fitness, or grammatical coherence, of the population, that is, the probability that a sentence said by one person is understood by another person. This equation describes a mutation-selection process in a population of individuals of n types.

COHERENCE THRESHOLD IN POPULATION LEARNING

Numerical simulations (Komarova et al., 2001; Nowak et al., 2001) and analytical estimates (Komarova, 2004) of equation (1) show the following trend. If the matrix, $\hat{Q}$, is close to identity, there are many coexisting localized steady-state solutions (corresponding to stable fixed points). For each such solution, the majority of the population speaks one of the languages, and the grammatical coherence, ϕ, takes values close to 1. As $\hat{Q}$ deviates far from the identity matrix (which means that there is a lot of "noise" in the system, that is, mistakes or learning are very likely), then this localization is lost and grammatical coherence becomes low.

A particular, highly symmetrical case of this system has been analyzed by Komarova et al. (2001) and Mitchener (2003). It was found that the low-coherence delocalized solution undergoes a transcritical bifurcation for the value $\Delta Q = || -\hat{I}||$ where $\hat{I}$ is the identity matrix defined by the entries of the matrix $\hat{A}$. A very interesting fact is that the threshold value of ΔQ does not depend on the dimension of the system, n.

A natural question is then to describe the phenomenon of the loss/gain of coherence for general matrices $\hat{A}$ and $\hat{Q}$. For instance, we can assume that the entries of the matrix $\hat{A}$ are taken from a distribution, and the matrix $\hat{Q}$ is a function of $\hat{A}$. Our results (Komarova, 2004) suggest that the threshold value of ΔQ typically tends to a constant as $n \to \infty$, where n is the size of the system. This finding can be called a universality property of universal grammars. Thus, for a (reasonable) class of learning algorithms (matrix $\hat{Q}$) and for any size of universal grammar, n, there is a finite coherence threshold in the system defined by the similarity of the grammars (matrix $\hat{A}$).

What is the significance of the coherence threshold for our understanding of language evolution? Our analyses can help us obtain possible bounds on complexity of universal grammar that are compatible with Darwinian evolution. Indeed, if the space of all possible grammars is too large, learning would take too long (humans have a limited time for learning before they become adults). At this point, linguistics meets evolutionary biology—there is a selection pressure to make universal grammar smaller and easier to learn. However, a larger pool of grammars also has some advantages, such as increased flexibility and more likely innovation.

DISCUSSION

There are two common misconceptions about the evolution of language. The first one represents the human capacity for language as an indivisible unit and implies that its gradual evolution is impossible because no component of this unit would have any function in the absence of the other components. For example, syntax could not have evolved without phonology or semantics and vice versa. The other misconception is that evolution of language started from scratch some 5 million years ago, when humans and chimps diverged. There are virtually no data to support this theory.

Both of these views are fundamentally flawed. First, all complex biological systems consist of specific components, although it is often hard to imagine the usefulness of individual components in the absence of other components. The usual task of evolutionary biology is to understand how complex systems can arise from simpler ones gradually by mutation and natural selection. In this sense, human language is no different from other complex traits.

Second, it is clear that the human language faculty did not evolve *de novo* in the last few million years, but was a continuation of a process that had evolved in other animals over a much longer time. Many animal species have sophisticated cognitive abilities in terms of understanding the world and interacting with one another. Furthermore, it is a well known "trick" of evolution to use existing structures for new, sometimes surprising purposes. Monkeys, for example, appear to have brain areas similar to our language centers, but they use them to control facial muscles and analyze auditory input. It may have been an easy evolutionary task to reconnect these centers for human language. Hence, the human language instinct is most likely not the result of a sudden moment of inspiration of evolution's blind watchmaker, but rather the consequence of several hundred million years of "experimenting" with animal cognition.

The goal of this paper is to show how methods of formal language theory, learning theory, and evolutionary biology can be combined to improve our understanding of the origins and properties of human language. We have formulated a mathematical theory for the population dynamics of grammar acquisition. The key result here is a "coherence threshold" that relates the maximum complexity of the search space to the amount of linguistic input available to the learner and the performance of the learning procedure. The coherence threshold represents an evolutionary stability condition for the language acquisition device. Only a universal grammar that operates above the coherence threshold can induce and maintain coherent communication in a population.

ACKNOWLEDGMENTS

Support from the Alfred P. Sloan foundation is gratefully acknowledged.

REFERENCES

Aitchinson, J. 1996. The Seeds of Speech. New York: Cambridge University Press.

Batali, J. 1994. Innate Biases and Critical Periods. Pp. 160–171 in Artificial Life IV, edited by R. Brooks and P. Maes. Cambridge, Mass.: MIT Press.

Bickerton, D. 1990. Language and Species. Chicago, Ill.: University of Chicago Press.

Cangelosi, A. 2001. Evolution of communication and language using signals, symbols and words. IEEE Transactions on Evolutionary Computation 5(2): 93–101.

Cangelosi, A., and D. Parisi, eds. 2001. Simulating the Evolution of Language. New York: Springer Verlag.

Chomsky, N.A. 1956. Three models for the description of language. IRE Transactions in Information Theory 2(3): 113–124.

Chomsky, N.A. 1957. Syntactic Structures. New York: Mouton.

Chomsky, N.A. 1965. Aspects of the Theory of Syntax. Cambridge, Mass.: MIT Press.

Chomsky, N.A. 1972. Language and Mind. New York: Harcourt Brace Jovanovich.

Christiansen, M.H., and S. Kirby, eds. 2003. Language Evolution: The States of the Art. New York: Oxford University Press.

Deacon, T. 1997. The Symbolic Species. London, U.K.: Penguin Books.

Dunbar, R. 1996. Grooming, Gossip, and the Evolution of Language. Cambridge, U.K.: Cambridge University Press.

Eigen, M., and P. Schuster. 1979. The Hypercycle: A Principle of Natural Self-Organisation. Berlin: Springer Verlag.

Fitch, W.T. 2000. The evolution of speech: a comparative review. Trends in Cognitive Science 4(7): 258–267.

Gopnik, M., and M. Crago. 1991. Familial aggregation of a developmental language disorder. Cognition 39(1): 1–50.

Harrison, M.A. 1978. Introduction to Formal Language Theory. Reading, Mass.: Addison-Wesley.

Hauser, M.D. 1996. The Evolution of Communication. Cambridge, Mass.: Harvard University Press.

Hawkins, J.A., and M. Gell-Mann. 1992. The Evolution of Human Languages. Reading, Mass.: Addison-Wesley.

Hurford, J.R., and S. Kirby. 1998. Co-evolution of Language-Size and the Critical Period. Pp. 39–63 in New Perspectives on the Critical Period Hypothesis and Second Language Acquisition, edited by D. Birdsong. Mahwah, N.J.: Lawrence Erlbaum Associates.

Hurford, J.R., M. Studdert-Kennedy, and C. Knight, eds. 1998. Approaches to the Evolution of Language. Cambridge, U.K.: Cambridge University Press.

Jackendoff, R. 1997. The Architecture of the Language Faculty. Cambridge, Mass.: MIT Press.

Jackendoff, R. 1999. Parallel constraint-based generative theories of language. Trends in Cognitive Science 3(10): 393–400.

Jackendoff, R. 2001. Foundations of Language. New York: Oxford University Press.

Knight, C., M. Studdert-Kennedy, and J. Hurford. 2000. The Evolutionary Emergence of Language: Social Function and the Origins of Linguistic Form. New York: Cambridge University Press.

Komarova, N.L. 2004. Replicator-mutator equation, universality property and population dynamics of learning. Journal of Theoretical Biology 230(2): 227–239.

Komarova, N.L., and M.A. Nowak. 2001a. Natural selection of the critical period for language acquisition. Proceedings: Biological Sciences 268(1472): 1189–1196.

Komarova, N.L., and M.A. Nowak. 2001b. The evolutionary dynamics of the lexical matrix. Bulletin of Mathematical Biology 63(3): 451–484.

Komarova, N.L. and M.A. Nowak. 2001c. Population Dynamics of Grammar Acquisition. Pp. 149–164 in Simulating the Evolution of Language, edited by A. Cangelosi and D. Parisi. London: Springer Verlag.

Komarova, N.L., and I. Rivin. 2003. Harmonic mean, random polynomials and stochastic matrices. Advances in Applied Mathematics 31(2): 501–526.

Komarova, N.L., P. Niyogi, and M.A. Nowak. 2001. The evolutionary dynamics of grammar acquisition. Journal of Theoretical Biology 209(1): 43–59.

Lai, C.S., S.E. Fisher, J.A. Hurst, F. Vargha-Khadem, and A.P. Monaco. 2001. A forkhead-domain gene is mutated in a severe speech and language disorder. Nature 413(6855): 519–523.

Levin, S. 2002. Complex adaptive systems: exploring the known, the unknown and the unknowable. Bulletin of the American Mathematical Society 40(1): 3–19.

Lieberman, P. 1984. The Biology and Evolution of Language. Cambridge, Mass.: Harvard University Press.

Lieberman, P. 1991. On the Evolutionary Biology of Speech and Syntax. Pp. 409–429 in Language Origin: A Multidisciplinary Approach, edited by B. Bichakjian, A. Nocentini, and B. Chiareli. Dordrecht, The Netherlands: Kluwer.

Maynard Smith, J., and E. Szathmary. 1995. The Major Transitions in Evolution. New York: Oxford University Press.

Mitchener, W.G. 2003. Bifurcation analysis of the fully symmetric language dynamical equation. Journal of Mathematical Biology 46(3): 265–285.

Nowak, M.A., and N.L. Komarova. 2001. Towards an evolutionary theory of language. Trends in Cognitive Sciences 5(7): 288–295.

Nowak, M.A., N.L. Komarova, and P. Niyogi. 2001. Evolution of universal grammar. Science 291(5501): 114–118.

Nowak, M.A., N.L. Komarova, and P. Niyogi. 2002. Computational and evolutionary aspects of language. Nature 417(6889): 611–617.

Oliphant, M. 1999. The learning barrier: moving from innate to learned systems of communication. Adaptive Behavior 7(3/4): 371–384.

Pinker, S., and A. Bloom. 1990. Natural language and natural selection. Behavioral and Brain Sciences 13(4): 707–784.

Smith, K. 2002. The cultural evolution of communication in a population of neural networks. Connection Science 14(1): 65–84.

Steels, L. 2000. Language as a Complex Adaptive System. Pp. 17–26 in Parallel Problem Solving from Nature. PPSN-VI, edited by M. Schoenauer, K. Deb, G. Rudolph, X. Yao, E. Lutton, J.J. Merelo, and H.-P. Schwefel. Lecture Notes in Computer Science 2000. New York: Springer Verlag.

Steels, L. 2001. Language Games for Autonomous Robots. Pp. 16–22 in IEEE Intelligent Systems, Vol. 16, edited by N. Shadbolt. New York: IEEE Press.

Steels, L., and F. Kaplan. 1998. Spontaneous Lexicon Change. Pp. 1243–1249 in Proceedings of COLING-ACL. Montreal: Association for Computational Linguistics.

Vargha-Khadem, F., K.E. Watkins, C.J. Price, J. Ashburner, K.J. Alcock, A. Connelly, R.S.J. Frackowiak, K.J. Friston, M.E. Pembrey, M. Mishkin, D.G. Gadian, and R.E. Passingham. 1998. Neural basis of an inherited speech and language disorder. Proceedings of the National Academy of Sciences 95(21): 12695–12700.

Wexler, K., and P. Culicover. 1980. Formal Principles of Language Acquisition. Cambridge, Mass.: MIT Press.

Agent-Based Modeling as a Decision-Making Tool

ZOLTÁN TOROCZKAI
Los Alamos National Laboratory
Los Alamos, New Mexico

STEPHEN EUBANK
Virginia Polytechnic Institute and State University
Blacksburg, Virginia

Researchers have made considerable advances in the quantitative characterization, understanding, and control of nonliving systems. We are rather familiar with physical and chemical systems, ranging from elementary particles, atoms, and molecules to proteins, polymers, fluids, and solids. These systems have interacting particles and well defined physical interactions, and their properties can be described by the known laws of physics and chemistry. Most important, given the same initial conditions, their behavior is *reproducible* (at least statistically).

However, other types of ubiquitous systems are all around us, namely systems that involve living entities (i.e., agents) about which we have hardly any quantitative understanding, either on an individual or collective level. In this paper, we refer to collectives of living entities as "agent-based systems" or "agent systems" to distinguish them from classical particle systems of inanimate objects. Although intense efforts have been made to study these systems, no generally accepted unifying framework has been found. Nevertheless, understanding, and ultimately controlling the behavior of agent systems, which have applications from biology to the social and political sciences to economics, is extremely important. Ultimately, a quantitative understanding can be a basis for designing agent systems, like robots or rovers that can perform tasks collectively that would be prohibitive for humans. Examples include deep-water rescue missions, minefield mapping, distributed sensor networks (for civil and military uses), and rovers for extraterrestrial exploration.

Even though there is no unifying understanding of agent systems, some control over their behavior can be achieved via *agent-based modeling* tools. The idea behind agent-based modeling is rather simple—build a computer model of the agent system under observation using a bottom-up approach by trying to mimic as much detail as possible. This can be rather expensive, however, because it requires (1) data collection, (2) model building, (3) exploitation of the model and the collection of statistics, and (4) validation, which normally means comparing the output of the model with additional observations of the real system.

The agent models described in this paper took about nine years to develop at Los Alamos National Laboratory. However, the framework for these models can be used to simulate many similar circumstances and to make predictions.

PROPERTIES OF AGENT SYSTEMS

Agent-based systems are hard to describe and understand within a unified approach because they differ from classical particle systems in at least two ways. First, an agent is a complex entity that cannot be represented by a simple function, such as a Hamiltonian function of a classical system (e.g., a spin system). Second, the interaction topology, namely the rules by which particles interact with each other, is generally represented by a complex, *dynamic* graph (network), unlike the regular lattices of crystalline solids or the continuous spaces of fluid dynamics. In many cases, the notion of "locality" itself is elusive in agent-based networks; in social networks, for example, the physical or spatial locality of agents may have little to do with social "distances" and interactions among them. To illustrate the complex structure of a "particle," or agent, and its consequences we can use traffic, namely people (agents) driving on a highway, as an example. Keep in mind, however, that the statements in this description are generally applicable to other agent systems.

Agents have the following qualities:

- A set of variables, x, describes the *state of the agent* (e.g., position on the road, speed, health of the driver, etc.). The corresponding state space is X.
- A set of variables, z, describes the *perceived state of the environment, Z,* which includes other agents, if there are any (e.g., level of congestion, state/quality of the road, weather conditions, etc.).
- There is a set of allowable *actions* (output space), A (swerve, brake, accelerate, etc.).
- A set of *strategies*, which are functions, $s\colon (Z \times X) \rightarrow A$, summon an action to a given circumstance, current state of the agent, and history up to time, t. These are "ways of reasoning" for the agent. One might think of strategies as *behavioral input space*. For example, depending on age, background, and other

factors, some drivers will brake and some will swerve to avoid an accident. Social studies and surveys can supply valuable statistical inputs, such as data showing that agents with n years of driving experience between ages a_1 and a_2 swerve f percent of the time and break g percent of the time.

- There is a set of *utility variables,* $u \in U$ (e.g., time to destination, number of accidents, number of speeding tickets, etc.).
- A *multivariate objective function,* $F{:}U \rightarrow R^m$, might include constraints ("rules") (e.g., the agent has to stay on the road). The analogous version in physics is called action. The agent *tries to optimize* this objective function (e.g., by minimizing the time to destination, avoiding accidents, etc.).

Unlike particles in classical systems, agents usually have *memory*, which they can use to change/evolve strategies, a process called *learning*. Another important aspect of agents is that they can *reason and plan*, which entail searching the choice tree and assigning weights and payoffs in light of what other agents might choose. In realistic situations that involve hundreds of agents (such as markets or traffic), long-term planning and reasoning are impossible because of the combinatorial explosion of possibilities and also because not all of the information is available to any single agent. Therefore, agents try to identify and exploit patterns in the responses of the surrounding environment to past actions and use these patterns to discriminate among strategies, reinforcing some and diminishing others (*reinforcement learning*). This leads to bounded-rationality-like behavior and introduces de-correlations between strategies; for that reason, reinforcement learning actually makes statistical modeling plausible.

In the following sections, we briefly describe two large-scale agent-based models developed at Los Alamos National Laboratory, a traffic simulator (TRANSIMS) and an epidemics simulator (EPISIMS).

TRANSIMS

The transportation analysis and simulation system (TRANSIMS) is an agent-based model of traffic in a particular urban area (the first model was for Portland, Oregon). TRANSIMS conceptually decomposes the transportation planning task using three different timescales. A large time scale associated with land use and demographic distribution (Figure 1) was used to create activity categories for travelers (e.g., work, shopping, entertainment, school, etc.). Activity information typically consisted of (1) requests that travelers be at a certain location at a specified time and (2) information on travel modes available to the traveler. For the large timescale, a synthetic population was created and endowed with demographics matching the joint distributions in census data. Synthetic households were also created based on survey data from several thousand households that included observations of daily activity patterns for each individual in the household. These activity patterns were then associated with syn-

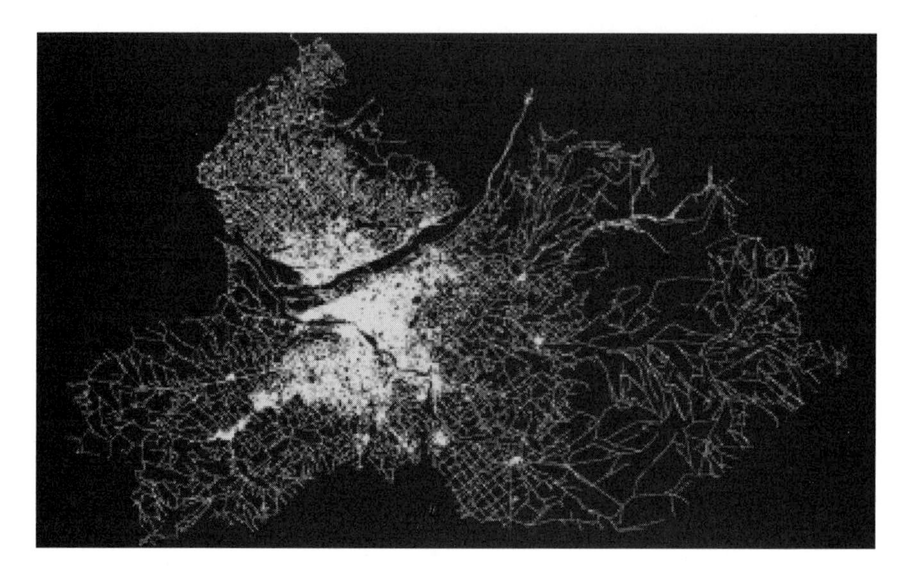

FIGURE 1 Large-scale view of the roadway network in Portland, Oregon.

thetic households with similar demographics. The locations of various activities were estimated, taking into account observed land-use patterns, travel times, and the dollar costs of transportation.

For the intermediate timescale (Figure 2), routes and trip chains were assigned to satisfy the activity requests. The estimated locations were fed into a routing algorithm to find minimum-cost paths through the transportation infrastructure consistent with constraints on mode choices (Barrett et al., 2001, 2002). For example, a constraint might be "walk to a transit stop, take transit to work using no more than two transfers and no more than one bus."

Finally, a very short time-scale (Figure 3) was used, associated with the actual execution of trip plans in the road network. This is done by a cellular automata simulation through a very detailed representation of the urban transportation network. The simulation, which resolved distances down to 7.5 meters and times down to 1 second, in effect resolved the traffic congestion caused by everyone trying to execute plans simultaneously by providing updated estimates of time-dependent travel times for each edge in the network, including the effects of congestion. These estimates were fed to a router and location estimation algorithms that produced new plans.

The feedback process continued iteratively until the system converged in a "quasi-steady state" in which no agent could find a better path in the context of every other agent's decisions. The resulting traffic patterns compared well to

FIGURE 2 Intermediate timescale view of locations and roads in downtown Portland, Oregon.

FIGURE 3 The TRANSIMS very-small timescale (microsimulation) of downtown Portland, Oregon.

observed traffic. Thus, the entire process estimated the demand on a transportation network using census data, land-usage data, and activity surveys.[1]

EPISIMS

Another application of the TRANSIMS model is in the field of epidemiology. Diseases, such as colds, flu, smallpox, and SARS, are transmitted through the air between two agents. These agents must either spend a long enough time in proximity to each other or be in a building with a closed air-ventilation system to transmit the disease. Thus, we can assume that the majority of infections take place in *locations*, like offices, shopping malls, entertainment centers, and mass transit units (metros, trams, etc.).

By tracking the people in our TRANSIMS virtual city, we generated a *bipartite contact network*, or graph, formed by two types of nodes—people nodes and location nodes. If a person, p, entered a location, l, an edge was drawn between that person and the corresponding location node on the graph. The edge had an associated time stamp representing the union of distinct time intervals the person, p, was at the location, l, during the day. If two people nodes, p_1 and p_2, had an incident edge in the same location node, l, the common intersection of the two time stamps told us the total time the two people spent in proximity to each other during the day, thus enabling us to determine the possibility of transmission of an airborne infection.

Using Portland as an example, there were about 1.6 million people nodes, 181,000 location nodes, and more than 6 million edges between them. This dynamic contact graph enabled us to simulate different disease-spread scenarios and test the sensitivities of epidemics to disease parameters, such as incubation period, person-to-person infection rates, influence of age structure, activity patterns, and so on. The epidemiological study tool thus generated, called EPISIMS, can be used as an aid to decision making and planning, for example, for an outbreak of smallpox.

Based on the Portland data, we arrived at the following findings for the spread of smallpox.[2] First, a person who has been vaccinated can be removed from the contact graph, along with his or her incident links. Thus, an efficient vaccination strategy removes the smallest subset of nodes, so that the resulting graph has many small disconnected pieces, which eliminates the spread of disease throughout the population. Unfortunately, the smallpox vaccine is not entirely harmless. In some people, it causes a disease called vaccinia that is some-

---

[1]More information, including availability of the software, can be obtained from *http://transims.tsasa.lanl.gov*.

[2]For more details on EPISIMS, see Eubank et al., 2004.

times fatal. Therefore, to minimize the incidence of vaccinia, mass vaccination (proposed by Kaplan et al., 2002) must be a last resort.

After studying the projection of the bipartite graph onto people nodes, however, we found very high expansion properties, and the only way to avoid mass spread of the disease would be to vaccinate everyone who had 10 or more contacts during the day, which effectively meant mass vaccination. Ultimately, we found that vaccinating people who frequently took long-range trips across the city, corresponding to shortcuts in the network (Watts and Strogatz, 1998), made it possible for us to use a more localized graph with a larger diameter. In case of an outbreak, this would allow for a ring strategy for quarantining and further vaccinations to stop the spread of disease.

Finally, the crucial parameter in containing an epidemic is the delay in reaction time. If we assume that sensors can perform an online analysis of pathogens in the air, the question is where they should be placed to be most effective, that is, where they would capture the onset of the outbreak. Due to a particular so-called scale-free property (Albert and Barabási, 2002) of the locations projection of the bipartite network, one can pinpoint a small set of locations (the so-called dominating set, about 10 percent of all locations) that would cover about 90 percent of the population and would thus be optimal locations for detectors. The same locations could be used for distribution purposes (e.g., of prophylactics and supplies). Figure 4 shows the evolution of epidemics after a covert introduction in a particular location (at a university) when the disease is left to spread (left side) compared to using a targeted-contact tracing and quarantining strategy (right).

ACKNOWLEDGMENTS

This work was done in collaboration with the TRANSIMS and EPISIMS teams. EPISIMS was supported by the National Infrastructure Simulation and Analysis (NISAC) Program at the U.S. Department of Homeland Security. The TRANSIMS project was funded by the U.S. Department of Transportation. The work of Zoltán Toroczkai was supported by the U.S. Department of Energy.

a.

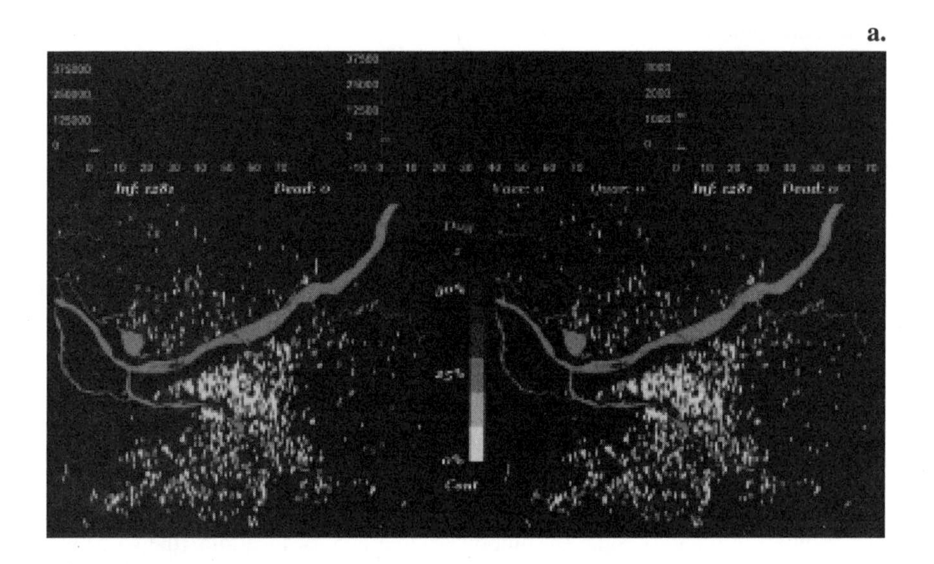

b,

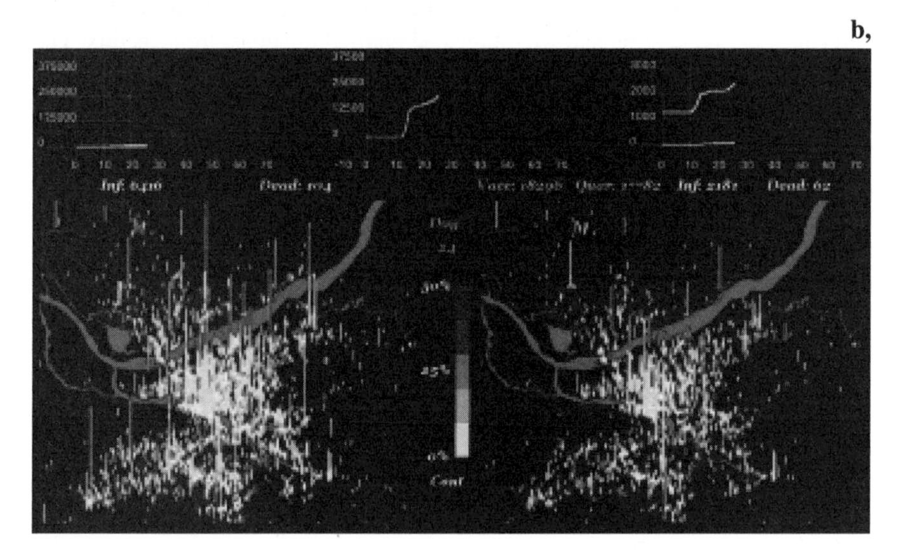

FIGURE 4 Comparison of baseline cases (on the left), with targeted (a.) vaccination and (b.) quarantining strategies (right). The bars represent the number infected at each location, and the light color represents the fraction of infected people who are infectious. The inserts show the cumulative number of people infected or dead as a function of time, and for the targeted response, the number vaccinated and quarantined. Note the different scales between the inserts. Source: LANL, 2005.

REFERENCES

Albert, R., and A.-L. Barabási. 2002. Statistical mechanics of complex networks. Review of Modern Physics 74(1): 47–97.

Barrett, C., R. Jacob, and M.V. Marathe. 2001. Formal language-constrained path problems. Journal on Computing 30(3): 809–837.

Barrett C., K. Bisset, R. Jacob, G. Konjevod, and M.V. Marathe. 2002. An Experimental Analysis of a Routing Algorithm for Realistic Transportation Networks. Technical Report No. LA-UR-02-2427. Proceedings of the European Symposium on Algorithms, Rome, Italy. Los Alamos, N.M.: Los Alamos National Laboratory.

Eubank, S., H. Guclu, V.S.A. Kumar, M.V. Marathe, A. Srinivasan, Z. Toroczkai, and N. Wang. 2004. Modelling disease outbreaks in realistic urban social networks. Nature 429(6988): 180–184.

Kaplan, E., D. Craft, and L. Wein. 2002. Emergency response to a smallpox attack: the case for mass vaccination. Proceedings of the National Academy of Sciences 99(16): 10935–10940.

LANL (Los Alamos National Laboratory). 2005. Controlling Smallpox: Strategies in a Virtual City Built from Empirical Data. Available online at: *http://episims.lanl.gov.*

Watts, D., and S. Strogatz. 1998. Collective dynamics of small-world networks. Nature 393(6684): 440–442.

ADDITIONAL READING

Barrett, C., S. Eubank, S.V. Anil Kumar, and M. Marathe. 2004. Understanding Large-Scale Social and Infrastructure Networks: A Simulation Based Approach. SIAM News, March 2004.

Barrett, C., S. Eubank, and J. Smith. 2005. If smallpox strikes Portland. Scientific American 292(3): 54–61.

Barrett, C., S. Eubank, and M. Marathe. Forthcoming. Modeling and Simulation of Large Biological, Information and Socio-Technical Systems: An Interaction-Based Approach. In Interactive Computation: The New Paradigm, edited by D. Goldin, S. Smolka, and P. Wegner. New York: Springer Verlag.

Halloran, M., I.M. Longini Jr., A. Nizam, and Y. Yang. 2002. Containing bioterrorist smallpox. Science 298(5597): 1428–1432.

Energy Resources for the Future

Introduction

ALLAN J. CONNOLLY
GE Energy
Schenectady, New York

JOHN M. VOHS
University of Pennsylvania
Philadelphia, Pennsylvania

With growing concerns about dwindling reserves of fossil energy sources, especially oil, and the impact of using these reserves on the environment, energy efficiency, alternative energy resources, and a proposed hydrogen economy have all recently become important subjects of science and engineering research and public policy. Attaining the goal of using more sustainable and environmentally benign energy resources will clearly require the development of a variety of new technologies. The goal of this session is to provide an overview of policy and technological issues in the important fields of energy resources and energy conversion.

1. Energy for the Future

The first talk by John Reinker of GE Global Research, provides a starting point for a discussion of future sources of energy. Dr. Reinker presents an overview of the energy landscape—current and potential sources of energy and energy usage—and options for the future—the proposed hydrogen-based energy economy and renewable sources of energy.

2. Research and Development on Hydrogen Production and Storage

Sunita Satyapal of the U.S. Department of Energy (DOE) presents an overview of DOE's Hydrogen Fuel Initiative, which was established to accelerate research and development (R&D) on hydrogen and fuel cells. Dr. Sataypal describes key challenges in hydrogen production, especially from renewable sources (e.g., photobiological), and hydrogen storage. She also describes metal hydrides, chemical hydrides, and carbon-based materials for hydrogen storage and discusses the most promising options. Finally, she identifies the most significant issues that remain to be addressed.

3. Fuel Cells: Current Status and Future Challenges

Fuel cells, which convert chemical energy sources directly to electricity, could potentially provide more efficient energy conversion than we now have. Stuart Adler of the University of Washington focuses on the role of fuel cells in the broad spectrum of energy choices and the proposed hydrogen economy. After briefly reviewing the basics of polymer-electrolyte and solid-oxide fuel cells, Dr. Adler gives an overview of current R&D and describes the technological challenges that must be overcome before the widespread use of fuel cells will be feasible.

4. Advanced Photovoltaics

Photovoltaic (PV) cells can harvest the abundant supply of solar energy and generate electricity without releasing carbon into the atmosphere. Before PV cells can be used to provide a significant fraction of our energy needs, however, we must increase their efficiency and find low-cost production processes. Michael McGehee of Stanford University briefly compares the most promising materials approaches to making solar cells, including silicon; multijunction, bioinspired, and inorganic thin films; and organic materials. Dr. Adler then focuses on R&D on creating nanostructured organic or organic-inorganic hybrid solar cells at extremely low cost. He discusses the design of these cells and the physics of light absorption, exciton diffusion, electron transfer, and charge transport.

Future Energy

JOHN K. REINKER
GE Global Research
Niskayuna, New York

Abundant, low-cost energy is essential to a modern society and a necessity for economic expansion, and transportation, electricity, and industrial requirements necessitate a balance of energy sources. Beyond low cost, however, pressures are increasing for environmentally sound energy solutions. In this presentation, I describe the development of the current electrical energy status in the United States and then discuss leading-edge developments in technology that will transform the energy portfolio of the future and support continued growth.

HISTORICAL PERSPECTIVE

Early industry in the United States was dependent on water for both transportation and power. The advent of electricity, a clean energy carrier, broke the geographic tie between energy source and energy use. The combination of electricity and the development of the internal combustion engine at the turn of the 20th century accelerated economic development.

The electrical grid in the United States was also dependent on hydropower, and many of the earliest hydroelectric dams are still operating throughout the country. As the grid spread, major changes in waterways occurred as a result of hydroelectric power projects and many of the natural hydropower sources were utilized. Eventually, the amount of power was restricted by the lack of new hydropower locations. Thus, other technologies became necessary to meet the growing electrical needs of the country.

Tremendous increases in electric power were achieved through the harnessing of coal. Even though overall power-plant efficiencies were in the 10 percent range, steady improvements were made, such as improved materials that allowed for higher steam pressures and temperatures. Improvements in the Rankin cycle, including feed-water heaters and reheat cycles, led to efficiencies approaching 40 percent. Today, nominal 1,000°F, 3,500 psi steam conditions are common.

With the development of nuclear technologies in the mid-20th century, nuclear power plants became economically attractive. In the United States, more than 100 nuclear power plants were installed, and they now produce 16 percent of the electrical output. These light-water reactors provide energy that powers steam-turbine generators. In the aftermath of the Three Mile Island incident, however, there was a shift away from building new nuclear power plants in the United States. Other nations, however, continued to build nuclear plants, and France now produces approximately 80 percent of its electricity from nuclear power. In the United States, nuclear power operators focused on improving plant performance and improved plant availability from 60 percent to more than 90 percent.

With the development of the jet engine, gas-fired turbines became available. By combining the combustion process and turbine technologies into a single package, a high power-density system was created. Gas turbines, which have rapid start-up capabilities and low cost, initially provided great peak-power capabilities. Subsequently, advances in combined-cycle technology, in which the waste heat of a gas turbine is captured through a heat-recovery steam generator and then converted to electricity through a steam-turbine generator, greatly increased the efficiency of power plants. A large number of plants that use this combined-cycle technology have been constructed in the past decade.

Recently, renewable technologies have been included in the utility purchase portfolio. Large blocks of wind power are being installed by major utilities, and solar panels are becoming cost effective in the southwestern United States.

The United States now produces approximately half of its electricity from coal, with the remaining half from natural gas, nuclear, and hydropower feedstocks (Figure 1). The energy sources for different regions of the world vary significantly, driven mainly by the availability of natural resources, such as coal and hydropower, but also by public acceptance of nuclear power. Figure 2 shows the overall electrical feedstock usage for the world—where traditional electrical feedstocks are used and where renewable sources are increasingly being used.

ECONOMIC DRIVERS

Energy investments are "large bets" placed by corporations on technology, fuel pricing, and customer demand. Even though many assume that the overall efficiency of a power plant is the primary economic driver, savings from effi-

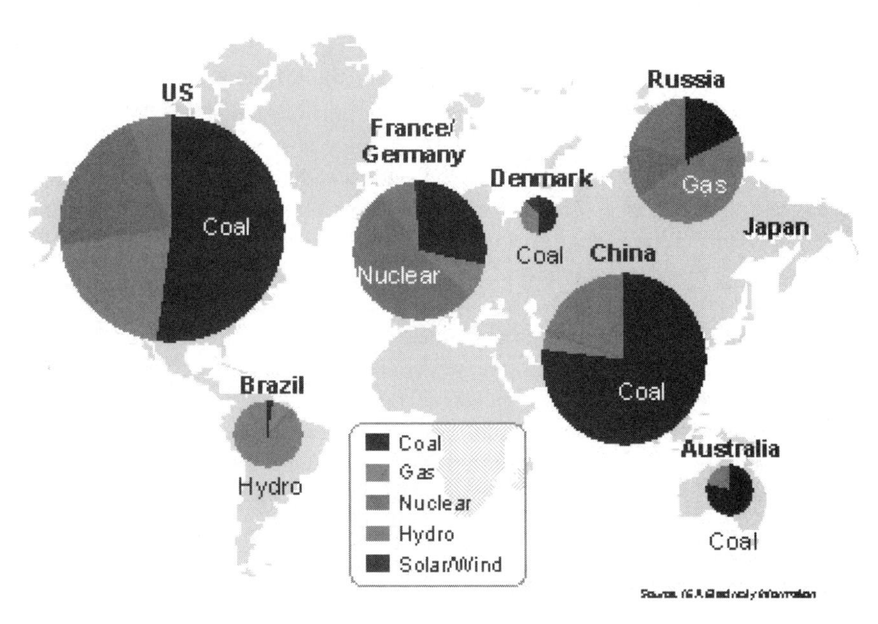

FIGURE 1 Energy feedstocks around the world. Data source: International Energy Agency Information Center, 2005.

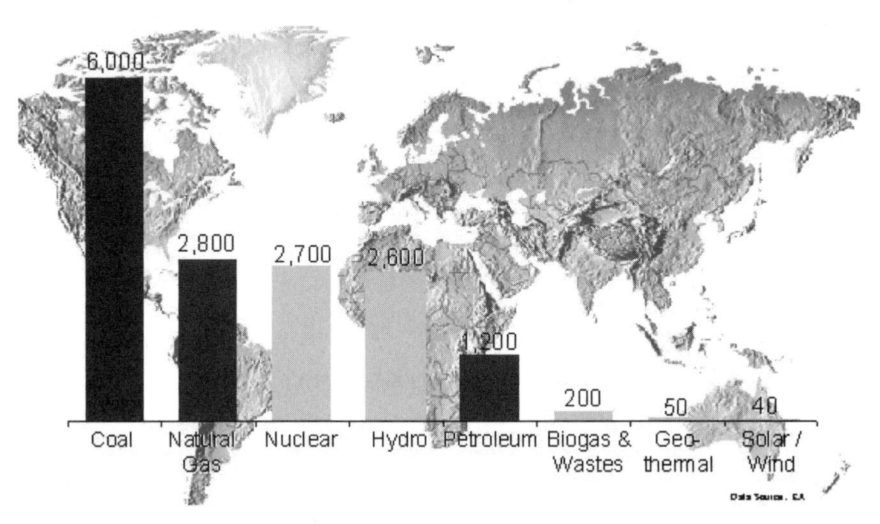

FIGURE 2 Comparison of energy sources worldwide in terawatt hours. Data source: International Energy Agency Information Center, 2005.

ciency may be offset by high fuel costs. For example, most of the natural-gas-fired, combined-cycle power plants commissioned during the recent U.S. power bubble represent the state of the art in terms of efficiency. Nevertheless, many of them have failed because of a spike in the price of natural gas. Even the most efficient plants could not overcome the cost penalty associated with high fuel costs and market-based pricing for electricity. Companies try to balance these risks by securing long-range fuel contracts and sales agreements. Companies are also required to manage emissions and may be required to pay taxes on carbon in the future.

To encourage the development of new technologies, governments can provide incentives for investments in certain energy sources. For example, in some regions of the United States, renewable portfolio mandates have been put in place. These mandates require that power from renewable sources be managed at the regional level so that the intermittency of wind and solar power do not affect the delivery of continuous energy to end customers.

EMERGENCE OF RENEWABLE POWER TECHNOLOGIES

Major advances have been made in wind turbine technologies. Most installed plants provide approximately 1 megawatt (MW) of power, but with the development of larger platforms, 5 MW offshore units are now being developed. These new plants will reduce capital investment and make wind power more competitive in the marketplace. The key technologies to making large units are longer blades, active wind controls, and power electronics to manage the conversion of the shaft torque to electrical power.

Solar power has a market advantage over other renewable power sources because it is competing in the residential market rather than the wholesale electricity market. The disadvantage of solar power is that the technology must be refined to reach the end consumer as part of an overall energy package. However, solar-power technology continues to improve, and today some homes are being built with integrated solar technology (Figure 3).

Technology development is under way on another attractive energy source, bioenergy (gases from landfills and coal that can be harnessed to generate electricity). Research is also under way on using agricultural feedstocks to generate electricity.

RETURN TO COAL AND NUCLEAR POWER

As the price of natural gas rises, the availability of low-cost coal and the strong performance of nuclear power plants have made traditional coal and nuclear plants more attractive. To make them more acceptable to the industry, regulators, and the public, many improvements are being developed.

A clear technological "game changer" for the coal industry is the develop-

FIGURE 3 A wind-turbine farm and a solar-powered home.

ment of advanced coal-gasification technologies. Conventional coal plants pulverize coal and then burn it in traditional boilers. Emissions must be cleaned up after combustion, and power is generated through a conventional Rankin cycle. With a coal gasifier, integrated coal combined-cycle (ICCC) plants can transform coal into a syngas, which enables the removal of many contaminants prior to combustion. In addition, the creation of the syngas means the Rankin cycle can be replaced with the highly efficient combined cycle.

In the long term, improved performance and lower emissions can also be achieved with advances in gas-turbine technologies. In the past 20 years, as a result of improved materials, coatings, and heat-transfer/cooling technologies, firing temperatures have increased from 2,000°F to 2,600°F. This has increased overall cycle efficiency from 50 percent to 60 percent, and improved emissions by an order of magnitude. Other manufacturers have improved performance by adding a second combustion system to improve overall efficiency at lower firing temperatures. In the future, solid-oxide fuel cells (SOFCs) could replace the combustion system of the gas turbine, resulting in a hybrid SOFC that could increase overall efficiencies to 70 percent or more.

Similar advancements are currently being made in the nuclear industry. Major reactor manufacturers in the United States are currently applying for licensing for new plants that will have larger power ratings than older plants and lower operating and maintenance costs per kW. In addition, these plants will have inherently safer designs with advanced control technologies, which will result in increased plant availability. Regulatory improvements have also been made: owners can apply for combined licenses instead of having to build plants under single-construction licenses and then apply for operating licenses.

FUTURE ENERGY

In the future, as energy demands rise worldwide, large quantities of power will be required that do not result in emissions of carbon dioxide. However, then, as now, no single energy solution will be right for any region. It is clear that we will be living in a carbon-constrained world in the future, and renewable sources will be part of the solution. In fact, there will be a portfolio of solutions, including nuclear power. Advances in technologies will continue to be critical.

Clear regulatory frameworks will be necessary to assure industries that they will have an adequate return on investments in technologies necessary to improve efficiencies and control emissions/greenhouse gasses. Energy equipment manufacturers will invest in a portfolio of solutions to meet their customers' needs. Satisfying the world's demand for energy will require a balanced portfolio of energy options, including coal, natural gas, nuclear, wind, hydropower, solar, and, eventually, bioenergy.

REFERENCE

International Energy Agency Information Center. 2005. Available online at: *http://www.iea.org/Textbase/subjectqueries/index.asp.*

Organic Semiconductors for Low-Cost Solar Cells

MICHAEL D. MCGEHEE
Stanford University
Stanford, California

CHIATZUN GOH
Stanford University
Stanford, California

Currently the world consumes an average of 13 terawatts (TW) of power. By the year 2050, as the population increases and the standard of living in developing countries improves, this amount is likely to increase to 30 TW. If this power is provided by burning fossil fuels, the concentration of carbon dioxide in the atmosphere will more than double, causing substantial global warming, along with many other undesirable consequences. Therefore, one of the most important challenges facing engineers is finding a way to provide the world with 30 TW of power without releasing carbon into the atmosphere. Although it is possible that this could be done by using carbon sequestration along with fossil fuels or by greatly expanding nuclear power plants, it is clearly desirable that we develop renewable sources of energy. The sun deposits 120,000 TW of radiation on the surface of the earth, so there is clearly enough power available if an efficient means of harvesting solar energy can be developed.

Only a very small fraction of power today is generated by solar cells, which convert solar energy into electricity, because they are too expensive (Lewis and Crabtree, 2005). More than 95 percent of the solar cells in use today are made of crystalline silicon (c-Si). The efficiency of the most common panels is approximately 10 percent, and the cost is $350/m^2$. In other words, the cost of the panels is $3.50/W of electricity produced in peak sunlight. When you add in the cost of installation, panel support, wiring, and DC to AC converters, the price rises to approximately $6/W. Over the lifetime of a panel (approximately 30 years), the average cost of the electricity generated is $0.30/kW-hr. By comparison, in most parts of the United States, electricity costs about \$0.06/kW-hr. Thus, it costs approximately five times as much for electricity from solar cells. If the cost of producing solar cells could be reduced by a factor of 10, solar energy would be not only environmentally favorable, but also economically favorable.

Although c-Si solar cells will naturally become cheaper as economies of scale are realized, dicing and polishing wafers will always be somewhat expensive. Thus, it is desirable that we find a cheaper way to make solar cells. The ideal method of manufacturing would be depositing patterned electrodes and semiconductors on rolls of plastic or metal in roll-to-roll coating machines, similar to those used to make photographic film or newspapers. Solar cells made this way would not only be cheaper, but could also be directly incorporated into roofing materials, thus reducing installation costs. Organic semiconductors that can be dissolved in common solvents and sprayed or printed onto substrates are very promising candidates for this application.

ORGANIC SEMICONDUCTORS

Because organic semiconductors have different bonding systems from conventional, inorganic semiconductors, they operate in a fundamentally different way. Conventional semiconductors are held together by strong covalent bonds that extend three-dimensionally, resulting in electronic bands that give rise to its semiconducting properties. Organic materials have similar intramolecular covalent bonds but are held together only by weak intermolecular van der Waals interactions. The electronic wave function is thus strongly localized to individual molecules, and the weak intermolecular interactions instigate a narrow electronic bandwidth formed in molecular solids.

Bonding

The semiconducting nature of organic semiconductors arises from the π electron bonds that exist when molecules are fully conjugated (i.e., have alternating single and double bonds). The weakly held π electrons are responsible for all interesting optical and electronic transitions in organic semiconductors. The π to π^* transitions in organic semiconductors are typically in the range of 1.4–2.5 eV, which overlaps well with the solar spectrum and makes them very promising candidates as active light absorbers in solar cells. A few examples of organic semiconductors used in solar cells are shown in Figure 1.

Excitons

The main difference between organic semiconductors and inorganic semiconductors as photovoltaic materials is that optical excitations of organic semiconductors create bound electron-hole pairs (called excitons) that are not effec-

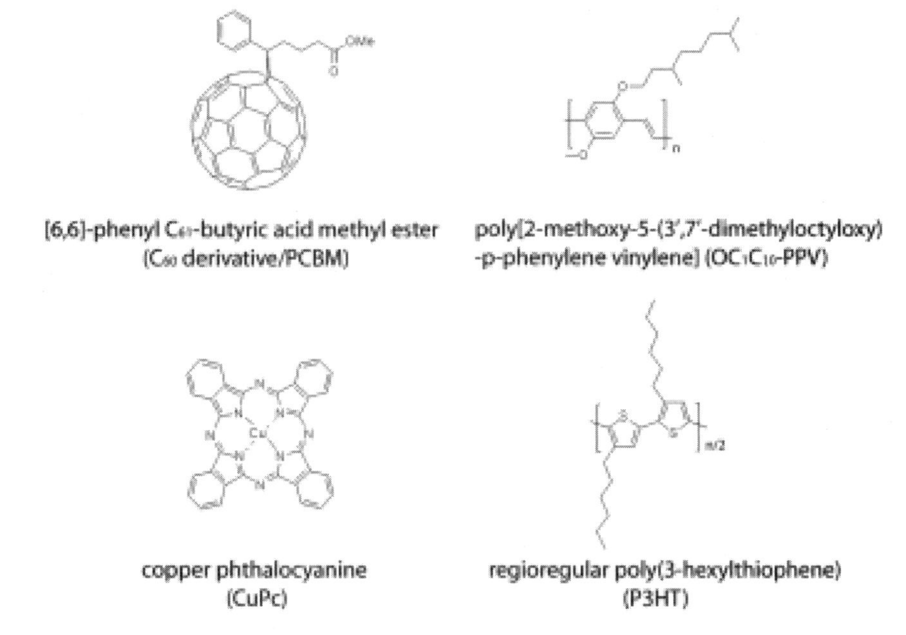

FIGURE 1 The chemical structures of four different organic semiconductors used in organic solar cells.

tively split by the electric field (Gregg, 2003). To separate the bound electrons and holes, there must be a driving force to overcome the exciton-binding energy, typically 0.1–0.4 eV. Excitons in organic semiconductors that are not split eventually recombine either radiatively or nonradiatively, thereby reducing the quantum efficiency of a solar cell.

In inorganic semiconductors the attraction between an electron-hole pair is less than the thermal energy kT. Therefore, no additional driving force is required to generate separated carriers. Research has shown that excitons in organic semiconductors can be efficiently split at a heterojunction of two materials with dissimilar electron affinities or ionization potentials.

Bandwidth

The narrow electronic bandwidth in organic semiconductors has a few consequences. First, the absorption-spectrum bandwidth is narrower than in conventional inorganic semiconductors. Consequently, a single organic material can be potentially photoactive only in a narrow optical-wavelength range of the solar spectrum (Figure 2). Although this is a disadvantage in terms of harvesting solar flux, multiple absorbers in stacks of solar cells connected in series can be engi-

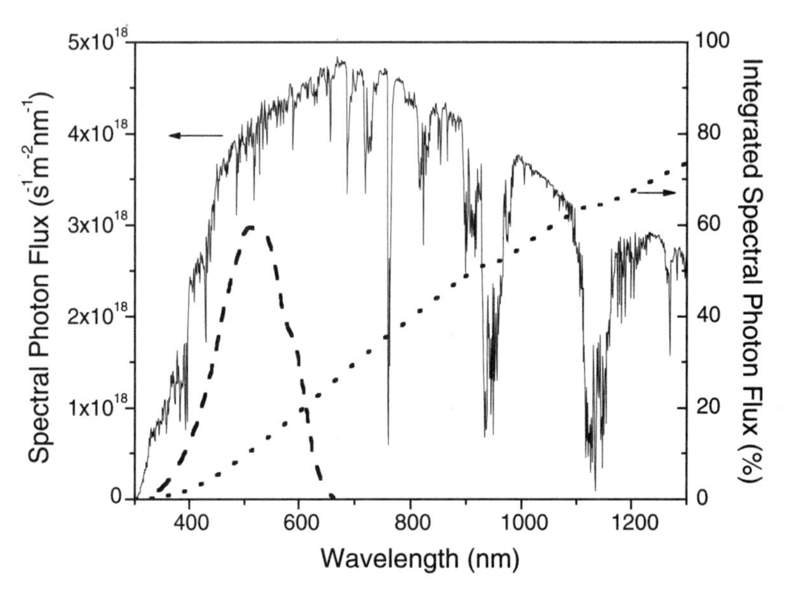

FIGURE 2 Overlap of the absorption spectrum of P3HT (dashed line) with the AM 1.5 solar spectrum (solid line). Integrated spectral-photon flux below each photon wavelength is also shown (dotted line).

neered to expand the absorption range. Because the valence band and conduction band are concentrated in narrower energy regions, the absorption coefficient resulting from the excitation of electrons from the valence band to the conduction band is very strong, typically $>10^{-5}$ cm^{-1} at peak absorption. This high absorption coefficient means that only a thin (100–200 nm) film is required to absorb most incident light, an attractive characteristic for solar cells because less material is required to make them.

Second, the charge carriers in organic semiconductors do not exhibit band-like transport as they do in inorganic semiconductors. Instead, they move around by a hopping mechanism between localized states. The charge-carrier mobilites in organic semiconductors are, therefore, inherently low, with typical values of $<10^{-2}$ cm^2/Vs. This low charge-carrier mobility puts a constraint on the thickness of organic materials that can be used in a solar cell because recombinative loss increases with increasing thickness. Fortunately, this drawback is offset because only a very thin layer of organic materials is necessary because organic semiconductors are highly absorptive. Organic solar cells may potentially perform better than conventional solar cells at higher temperature, because hopping is a thermally activated process. The performance of inorganic solar cells typically decreases as operating temperature increases.

A third key difference between organic and inorganic semiconductors is that organic materials do not have dangling bonds at surfaces. Therefore, organic-organic junctions or organic-metal junctions in organic solar cells (interface states) do not act as potential charge-carrier recombination sites.

PRODUCTION OF HETEROJUNCTION DEVICES

The simplest organic solar cells can be made by sandwiching thin films of organic semiconductors between two electrodes with different work functions. The work function is the amount of energy necessary to pull an electron from a material. When such a diode is made, electrons from the low-work-function metal flow to the high-work-function metal until the Fermi levels are equalized throughout the structure. This sets up a built-in electric field in the semiconductor. When the organic semiconductor absorbs light, electrons are created in the conduction band, and holes (positive-charge carriers) are created in the valence band. Thus, in principle, the built-in electric field can pull the photogenerated electrons to the low-work-function electrode and holes to the high-work-function electrode, thereby generating a current and voltage (Figure 3a). In practice, however, these cells have very low power-conversion efficiency (<0.1 percent) because the electric field is not strong enough to separate the bound excitons (i.e., the excited-state species formed in organic semiconductors described above).

A significant improvement in the performance of organic solar cells was achieved by Tang (1986). His device consisted of a heterojunction between donor and acceptor semiconductors, resembling a p-n junction in conventional solar cells (Figure 3b). The benefit of this device derived from the use of two organic materials with offset electron affinities (lowest unoccupied molecular orbital, LUMO) or ionization potentials (highest occupied molecular orbital, HOMO). Excitons that diffuse to the interface undergo efficient charge transfer, as this offset in the energy levels provides a sufficient chemical potential energy to overcome the intrinsic exciton-binding energy. Upon charge transfer, the electrons are transported in the acceptor material and the holes in the donor material to their respective electrodes.

The efficiency of this type of planar heterojunction device is limited, however, by the exciton diffusion length, which is the distance over which excitons travel before undergoing recombination, approximately 5–10 nm in most organic semiconductors. Excitons formed at a location further than 5–10 nm from the heterojunction cannot be harvested. The active area of this type of solar cell is thus limited to a very thin region close to the interface, which is not enough to adsorb most of the solar radiation flux.

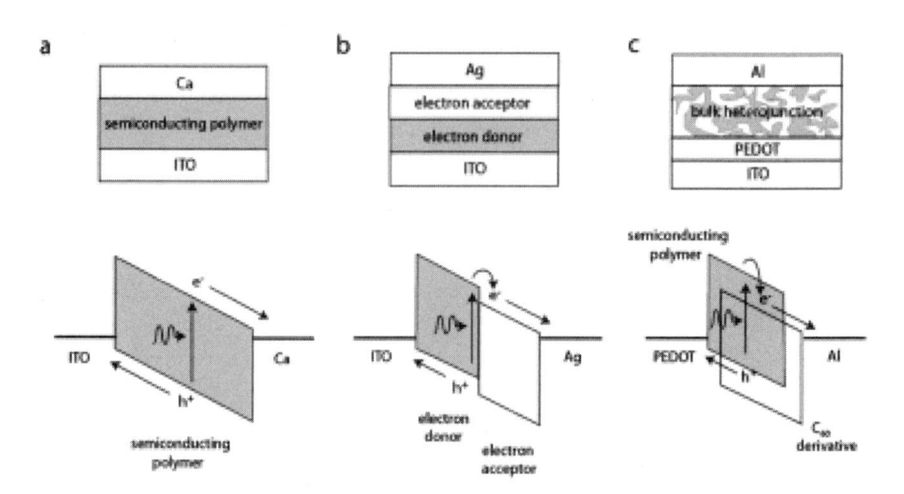

FIGURE 3 A schematic energy-band diagram of (a) a single-layer semiconductor polymer solar cell, in which indium-tin oxide (ITO) serves as a transparent high-work-function electrode and calcium serves as the low-work-function electrode; (b) photoinduced electron transfer from the lowest unoccupied molecular orbital (LUMO) of an electron donor to the LUMO of an electron acceptor in a planar heterojunction cell; (c) a bulk heterojunction solar cell based on semiconducting polymer and C_{60} derivative.

Blend Cells

In the mid 1990s, Yu and colleagues (1995) showed that excitons can be rapidly split by electron transfer before the electron and hole recombine if carbon-60 (C_{60}) derivatives are blended into the polymer (Figure 3c). Blend solar cells were made simply by blending the C_{60} derivative, which acts as an electron acceptor, into the polymer at concentrations in the range of 18–80 wt. percent. At these concentrations, the polymer and the C_{60} derivatives form a connected network to each electrode. The key to making efficient blend solar cells is to ensure that the two materials are intermixed very closely at a length scale less than the exciton diffusion length so that every exciton formed in the polymer can reach an interface with C_{60} to undergo charge transfer.

At the same time, the film morphology has to enable charge-carrier transport in the two different phases to minimize recombination. The film morphology (i.e., phase separation between the two materials) and, ultimately, the efficiency of the device, are determined by the concentration of materials, film-casting solvent, annealing time, temperature, and other parameters. Solar cells made by this method have continuously improved to better than 2 percent power efficiency under solar AM 1.5 conditions over the last few years (Padinger et al., 2003; Shaheen et al., 2001), and recently, an efficiency of 5 percent was reported (Ma et al., 2005).

The work on polymer/C_{60}-derivative blend cells has created a new paradigm in the field of organic-based solar cells, which is the notion of bulk heterojunction devices, wherein two semiconductors with offset energy levels are interpenetrated at a very small length scale to create a high interface area for achieving high-efficiency devices. Since then, similar bulk heterojunction devices using electron acceptors other than the C_{60} derivative, such as CdSe nanorods (Huynh et al., 2002), a second semiconducting polymer (Granstrom et al., 1998), and titania nanocrystals (Arango et al., 1999), have been demonstrated, albeit with slightly lower efficiencies.

Limits on Performance

To understand the limits on the performance of bulk heterojunction devices and find ways to improve them, it is important to consider all of the processes that must occur inside the cells for electricity to be generated. These processes, shown in Figure 4a, are: (1) light absorption; (2) exciton transport to the interface between the two semiconductors; (3) forward electron transfer; and (4) charge transport. One must also consider undesirable recombination processes that can limit the performance of the cell, such as geminate recombination of electrons and holes in the polymer and back electron transfer from the electron acceptor to the polymer (Figure 4b).

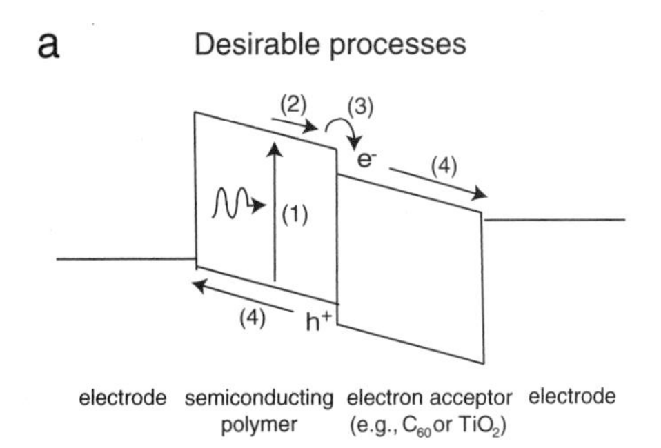

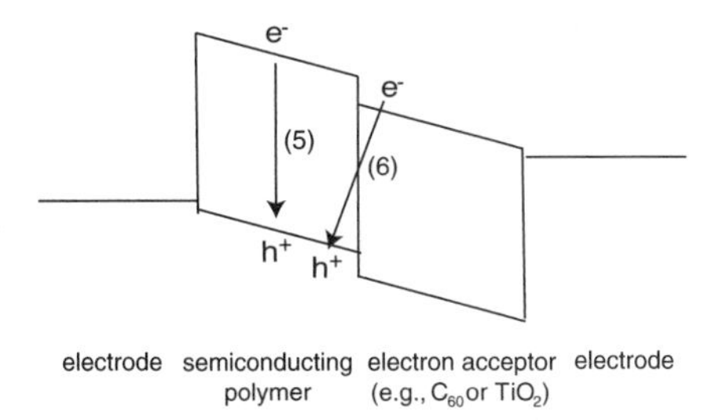

FIGURE 4 Schematic energy diagrams of the semiconductors and energy levels in a bulk heterojunction solar cell showing the (a) desirable and (b) undesirable processes that can occur.

Light Absorption

The necessity of absorbing most of the solar spectrum (process 1) creates two requirements. First, the band gap must be small enough to enable the polymer to absorb most of the light in the solar spectrum. Calculations to determine the band gap that optimizes the amount of light that can be absorbed and the voltage that can be generated show that the ideal band gap is approximately 1.5 eV, depending on the combination of semiconducting polymers and electron acceptors (Coakley and McGehee, 2004). Second, the film must be thick enough to absorb most of the light. For most organic semiconductors, this means that films must be 150–300 nm thick, depending on how much of the film consists of a nonabsorbing electron acceptor. The optimum film thickness will absorb much incident light without significant recombination losses.

Exciton Transport

Once an exciton is created in the polymer, it must diffuse or travel by resonance energy transfer (process 2) to the interface with the other semiconductor and be split by electron transfer before it recombines (process 5). Experiments have shown that an exciton can diffuse approximately 5–10 nm in most semiconducting polymers before recombination. Therefore, no regions in the polymer can be more than 5–10 nm from an interface. Templating or nanostructuring of the donor and acceptor phases to fabricate ordered bulk heterojunction with controlled dimensions is an attractive approach to achieving full exciton harvesting (Figure 5) (Coakley and McGehee, 2003). Some small-molecule semiconductors have been shown to have larger exciton diffusion lengths (Peumans et al., 2003).

Research is under way to improve exciton transport in organic semiconductors, for example by using resonance energy transfer to funnel excitons directly to an absorber located at the charge-splitting interface or by incorporating phosphorescent semiconductors, which exhibit longer excited-state lifetimes (Liu et al., 2005; Shao and Yang, 2005).

Forward Electron Transfer

The actual process of charge transfer (process 3) requires that the offset in LUMO levels of the donor and acceptor be sufficient to overcome the exciton-binding energy. However, this drop in energy must not be excessive, because the maximum voltage attainable from this type of bulk heterojunction solar cell is determined by the gap between the HOMO of the electron donor and the LUMO of the acceptor. The gap becomes smaller as the LUMO of electron acceptors is moved farther away from the LUMO of the polymer, which corresponds to a larger driving force for charge transfer.

As can be seen from processes 1, 2, and 3, the design of an efficient organic

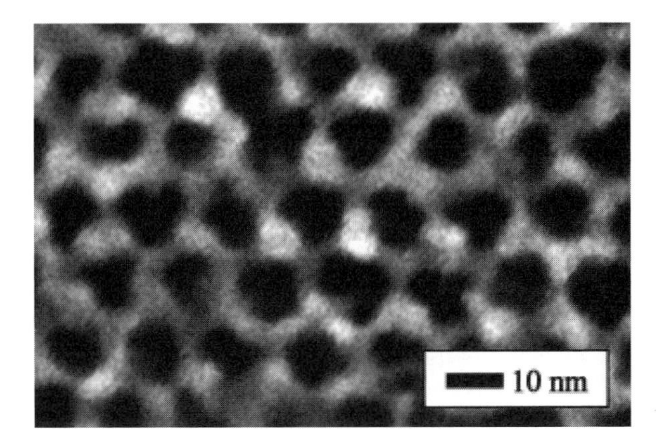

FIGURE 5 Scanning electron micrograph of mesoporous titania film. This film was used in combination with P3HT to fabricate ordered bulk heterojunction solar cells. Source: Coakley and McGehee, 2003. Copyright 2003, American Institute of Physics. Reprinted with permission.

solar cell involves optimizing the various energy levels to achieve the optimum level of extracted current with respectable voltage, as the power supplied by a solar cell is the product of current and voltage. Fortunately, the wealth of chemical synthetic knowledge and the dependence of electronic properties of organic molecules on their molecular structures allow for flexible tuning of the band gap and energy levels of organic semiconductors by chemical synthesis. Significant research on band engineering of this type should yield very promising results in the near future.

Charge Transport

After forward electron transfer, the holes in the polymer and the electrons in the electron acceptor must reach the electrodes (process 4) before the electrons in the acceptor undergo back electron transfer to the polymer (process 6). Even in the best bulk heterojunction cells, this competition limits the efficiency of the cells.

The problem can usually be mitigated by making cells that are only 100 nm thick so that the carriers do not have to travel very far. Unfortunately, most of the light is not absorbed by films this thin. If the films are thick enough to absorb most of the light, then only a small fraction of the carriers escape the device. Many researchers are now trying to optimize the interface between the two semiconductors and improve charge transport in the films so that the charge can be extracted from 300-nm-thick films before recombination occurs.

FUTURE CHALLENGES

The outlook for organic solar cells is very bright. Efficiency greater than 5 percent has been achieved (Ma et al., 2005; Xue et al., 2004), and many are optimistic that 20 percent can be achieved by optimizing the processes described above. Once this goal is achieved, a primary research challenge will be making cells that are stable in sunlight and that can handle wide temperature swings. The survival of many organic pigments in car paints in sunlight and the production of organic light-emitting diodes with operational lifetimes greater than 50,000 hours are encouraging signs that the required stability can be achieved. The final challenge will be scaling up the process and manufacturing the cells at a cost of approximately $30/m^2$. Shaheen and colleagues (2005) have described several approaches to making organic cells.

REFERENCES

Arango, A.C., S.A. Carter, and P.J. Brock. 1999. Charge transfer in photovoltaics consisting of polymer and TiO_2 nanoparticles. Applied Physics Letters 74(12): 1698–1700.

Coakley, K.M., and M.D. McGehee. 2003. Photovoltaic cells made from conjugated polymers infiltrated into mesoporous titania. Applied Physics Letters 83(16): 3380–3382.

Coakley, K.M., and M.D. McGehee. 2004. Conjugated polymer photovoltaic cells. Chemistry of Materials 16(23): 4533–4542.

Granstrom, M., K. Petritsch, A. Arias, A. Lux, M. Andersson, and R. Friend. 1998. Laminated fabrication of polymeric photovoltaic diodes. Nature 395(6699): 257–260.

Gregg, B. 2003. Excitonic solar cells. Journal of Physical Chemistry B 107(20): 4688–4698.

Huynh, W.U., J.J. Dittmer, and A.P. Alivisatos. 2002. Hybrid nanorod-polymer solar cells. Science 295(5564): 2425–2427.

Lewis, N.S., and G. Crabtree. 2005. Basic research needs for solar energy utilization. Available online at: *http://www.sc.doe.gov/bes/reports/files/SEU_rpt.pdf.*

Liu, Y., M.A. Summers, C. Edder, J.M.J. Fréchet, and M.D. McGehee. 2005. Using resonance energy transfer to improve exciton harvesting in organic-inorganic hybrid photovoltaic cells. Advanced Materials (in press).

Ma, W., C. Yang, X. Gong, K. Lee, and A.J. Heeger. 2005. Thermally stable, efficient polymer solar cells with nanoscale control of the interpenetrating network morphology. Advanced Functional Materials 15(10): 1617–1622.

Padinger, F., R.S. Rittberger, and N.S. Sariciftci. 2003. Effects of postproduction treatment on plastic solar cells. Advanced Functional Materials 13(1): 85–88.

Peumans, P., A. Yakimov, and S.R. Forrest. 2003. Small molecular weight organic thin-film photodetectors and solar cells. Journal of Applied Physics 93(7): 3693–3723.

Shaheen, S.E., C.J. Brabec, N.S. Sariciftci, F. Padinger, T. Fromherz, and J.C. Hummelen. 2001. 2.5% efficient organic plastic solar cells. Applied Physics Letters 78(6): 841–843.

Shaheen, S.E., D.S. Ginley, and G.E. Jabbour, editors. 2005. Organic Photovoltaics. MRS Bulletin 30(1). Special Edition.

Shao, Y., and Y. Yang. 2005. Efficient organic heterojunction photovoltaic cells based on triplet materials. Advanced Materials (in press).

Tang, C.W. 1986. Two-layer organic photovoltaic cell. Applied Physics Letters 48(2): 183–185.

Xue, J.G., S. Uchida, B.P. Rand, and S.R. Forrest. 2004. Asymmetric tandem organic photovoltaic cells with hybrid planar-mixed molecular heterojunctions. Applied Physics Letters 85(23): 5757–5759.

Yu, G., J. Gao, J.C. Hummelen, F. Wudl, and A.J. Heeger. 1995. Polymer photovoltaic cells: enhanced efficiencies via a network of internal donor-acceptor heterojunctions. Science 270(5243): 1789–1791.

Research and Development at the U.S. Department of Energy on Hydrogen Production and Storage

SUNITA SATYAPAL
U.S. Department of Energy
Washington, D.C.

Investigating the potential of hydrogen as an energy carrier is just one of many strategies in the U.S. Department of Energy (DOE) research and development (R&D) portfolio. Programs are under way on hydrogen, biomass, solar, wind, geothermal, and nuclear energy, as well as on improved use of fossil fuels, carbon sequestration, and advanced hybrid-vehicle technologies. In this presentation, I focus on the DOE Hydrogen Program and two critical areas of investigation—hydrogen production and hydrogen storage. Following a brief overview of hydrogen production strategies, I describe in detail the status and challenges of hydrogen storage.

U.S. petroleum dependence is driven by transportation, which accounts for two-thirds of the 20 million barrels of oil our nation uses each day (DOE, 2005a,b,c). Today, the United States imports 55 percent of its oil, and this is expected to increase to 68 percent by 2025 under a status quo scenario (DOE, 2005a,b,c). The public has few other options for transportation fuel because nearly all vehicles currently run on gasoline or diesel. To promote national energy security, we must develop alternative energy carriers.

Molecular hydrogen, the simplest diatomic molecule known, with the highest gravimetric energy content of known fuels, has the potential to be an attractive alternative energy carrier. Hydrogen could not only be clean and efficient, but it could also be derived from a variety of domestic resources, such as biomass, hydro, wind, solar, geothermal, and nuclear energy sources, as well as coal (with sequestration) and natural gas (for limited applications) in the near term

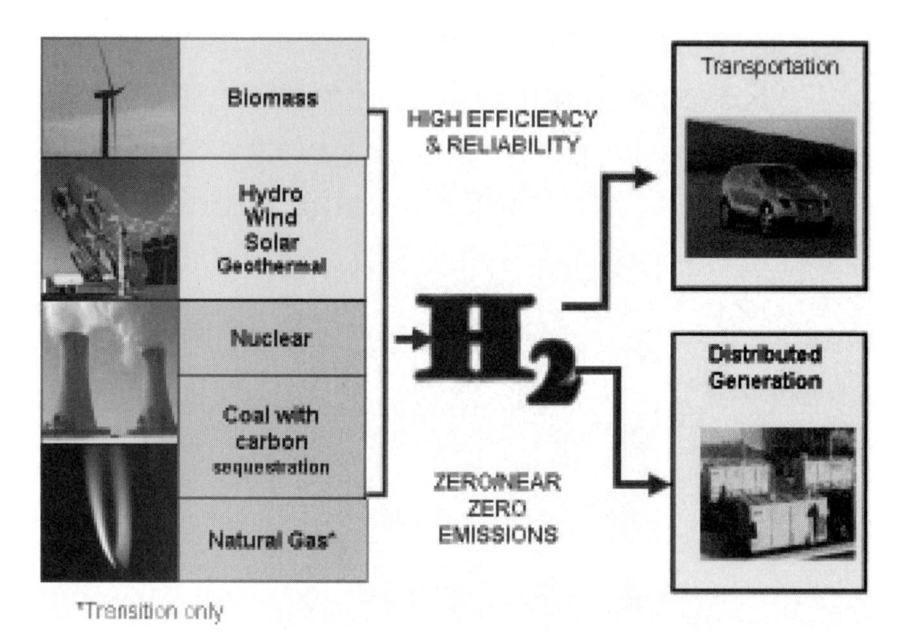

FIGURE 1 Hydrogen production strategy: To produce hydrogen from renewable, nuclear, and coal resources with technologies that will yield virtually zero criteria and greenhouse gas emissions. Source: U.S. DOE.

(see Figure 1). Hydrogen could then be used in high-efficiency power-generation systems, including internal combustion engines or fuel cells for both vehicular transportation and distributed electricity generation.

In addition to promoting energy security, pure hydrogen has the potential to be environmentally advantageous because the only by-products of hydrogen-powered fuel cells are water and heat. Emissions of carbon dioxide and criteria pollutants (e.g., nitrogen oxides [NO_x], sulfur oxides [SO_x], and carbon monoxide [CO], etc.) would essentially be eliminated from the point of use. These emissions would be easier to control at the point of generation, rather than from the tailpipes of 200 million vehicles.

However, although molecular hydrogen is abundant in the universe, it is not plentiful on Earth, and it is not a primary fuel source. The question is how can we efficiently produce and safely deliver, store, and use hydrogen to reap the benefits of reduced emissions, higher energy efficiency, and improved energy security.

HYDROGEN FUEL INITIATIVE

The Hydrogen Fuel Initiative, announced by President Bush in 2003, commits $1.2 billon over five years for accelerated R&D and demonstration programs that will support an industry decision in 2015 on the commercial viability of hydrogen and fuel-cell technologies. If industry decides to proceed, a full transition to a hydrogen economy would clearly take decades and strong government-industry partnerships. However, it is essential that R&D to address the feasibility of hydrogen occur now.

A brief overview of hydrogen production and the key technical challenges under investigation in the DOE Hydrogen Program will be followed by a discussion of the critical technical challenges of safe and efficient hydrogen storage. In 2003, DOE announced a "grand challenge" to the global technical community on hydrogen storage that culminated in the formation of the National Hydrogen Storage Project (see Figure 2). For the first time, centers of excellence (DOE, 2005a) dedicated to hydrogen storage were formed with multiple university, industry, and national laboratory partners, leveraging expertise and capabilities from all sectors to tackle this difficult problem. In addition, 17 new projects on the basic science of hydrogen storage were selected in 2005 and funded at $20 million over three years through the DOE Office of Science. The discussion that

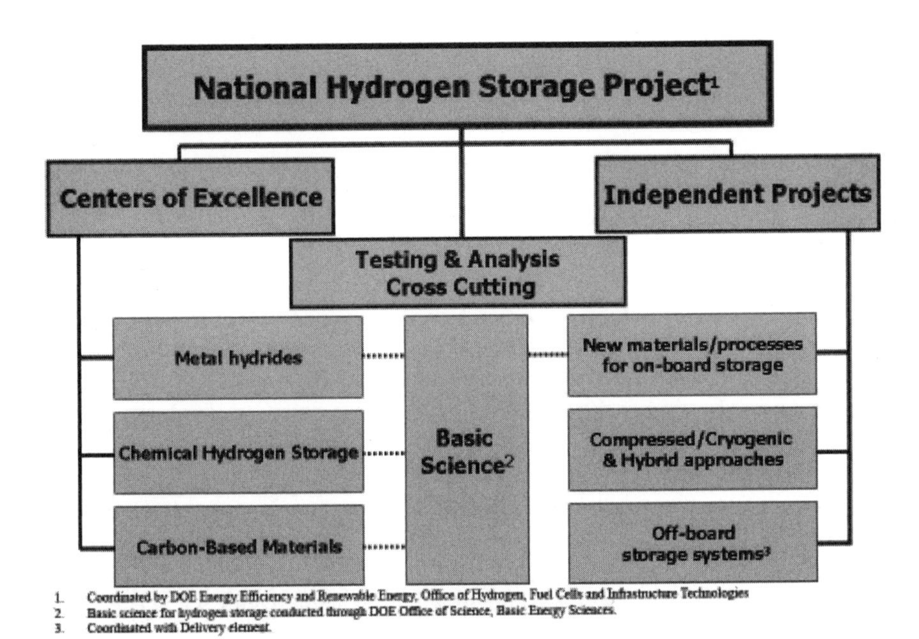

FIGURE 2 Based on the "Grand Challenge" launched by DOE in 2003, the National Hydrogen Storage Project was planned at roughly $150 million in federal funding over 5 years, subject to appropriation. An additional ~$20 million over 3 years is planned for 17 new projects selected by the Office of Science in 2005. Source: U.S. DOE.

follows includes a description of the key issues in hydrogen storage, the National Hydrogen Storage Project, and progress in the search for the "holy grail" of hydrogen storage.

SUMMARY OF RESEARCH AND DEVELOPMENT ON HYDROGEN PRODUCTION

To meet the needs of a hydrogen economy, ensure energy security, and realize environmental benefits, hydrogen must be produced from diverse resources with minimal life-cycle cost and environmental impact and with maximum energy efficiency. In the near term, to avoid large capital investments in infrastructure, small-scale distributed hydrogen production is likely, including distributed natural gas reforming or electrolysis at fueling stations. The key challenge is to meet the DOE goal of $2 to $3 per gallon gasoline equivalent (gge) by 2015. While hydrogen can, and will be, produced from different pathways and diverse resources, the $2 to $3 per gge goal must be met independent of production pathway or energy source. Recent results show that the cost of $5 per gge for hydrogen produced from natural gas in a distributed system (delivered, untaxed), (Devlin, 2005), is approaching $3 per gge.

To reduce cost further and improve durability and energy efficiency, research is under way on catalysts, membranes for separation and purification, water-gas shift reactors, and hydrogen-compression technology. Research on reforming of biomass and renewable liquids is addressing the same issues. The long-term strategy is to produce hydrogen from renewable sources, nuclear energy, and coal with sequestration (via gasification, not coal-based electricity) to achieve carbon-neutral or zero-carbon technologies.

Water electrolysis, another area of investigation, is focused on the key issues of capital cost and electricity cost. Research is under way on materials to improve electrolyzer durability and energy efficiency, and, at the same time, to reduce cost. Another approach under investigation involves high-temperature thermochemical reactions that can use heat from nuclear power plants or high-temperature solar energy. Key issues are efficiency, cost, and durability. A few of the other areas being studied are more robust materials for high-temperature operation, lower cost solar-concentrator technology, and optimized thermochemical reactions. Finally, exploratory research for long-term approaches is also under way on photobiological or photoelectrochemical production of hydrogen. In all of these areas, hydrogen purity is a key issue (e.g., 99.99 percent purity is necessary for polymer-electrolyte-membrane [PEM] fuel cells).

THE "GRAND CHALLENGE" OF HYDROGEN STORAGE

On the basis of weight, hydrogen has nearly three times the energy content of gasoline (120 MJ/kg for hydrogen versus 44 MJ/kg for gasoline). However,

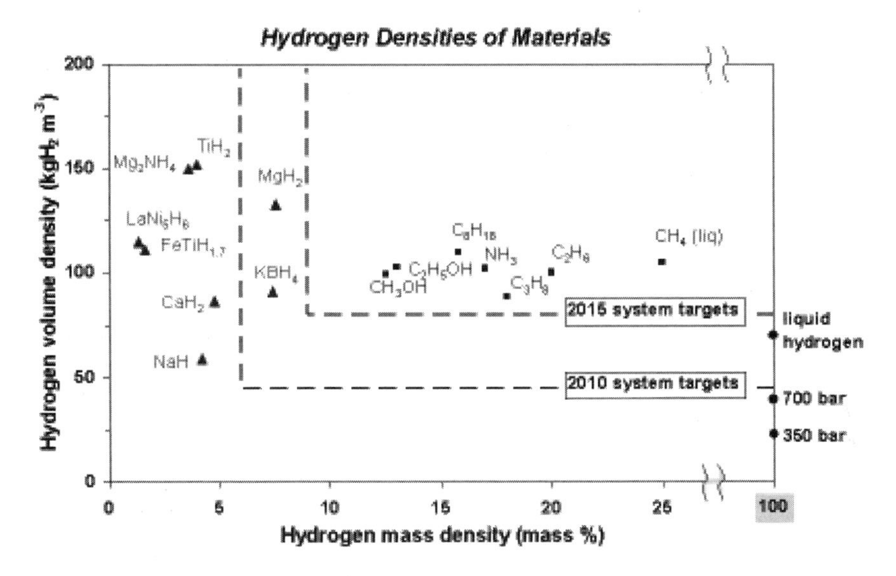

FIGURE 3 Examples of materials and hydrogen storage capacities. Source: U.S. DOE.

on the basis of volume, the situation is reversed (8 MJ/liter for liquid hydrogen versus 32 MJ/liter for gasoline). Onboard vehicular hydrogen storage is a critical challenge to meeting customer expectations for a driving range of more than 300 miles within the weight, volume, safety, and cost constraints of a marketable vehicle. Through the FreedomCAR and Fuel Partnership, between DOE and leading automotive and energy industries, technical targets were set for commercially viable vehicular hydrogen storage systems in the United States. Some of these system-level targets for 2010 are: gravimetric capacity of 6 weight percent (= 2.0 kWh/kg), volumetric capacity of 1.5 kWh/L (= 0.045 kg hydrogen/L) and a cost of $4/kWh (DOE, 2005d). Figure 3 shows various material capacities and total-system capacities for a limited number of systems, illustrating that both fundamental properties and system-engineering issues must be addressed to meet the targets (Ordaz et al., 2005).

Current hydrogen storage technologies include: high-pressure tanks, cryogenic storage, metal hydrides, chemical hydrides, and high-surface-area sorbents, such as nanostructured carbon-based materials. High-pressure and cryogenic tanks, high-surface-area sorbents, and many metal hydrides can be categorized as "reversible" onboard hydrogen storage because "refueling" with hydrogen can take place directly on board the vehicle. For chemical hydrogen storage and some high-temperature metal hydrides, hydrogen regeneration is not possible on board the vehicle; thus with these systems, hydrogen must be regenerated off board (see Figure 4).

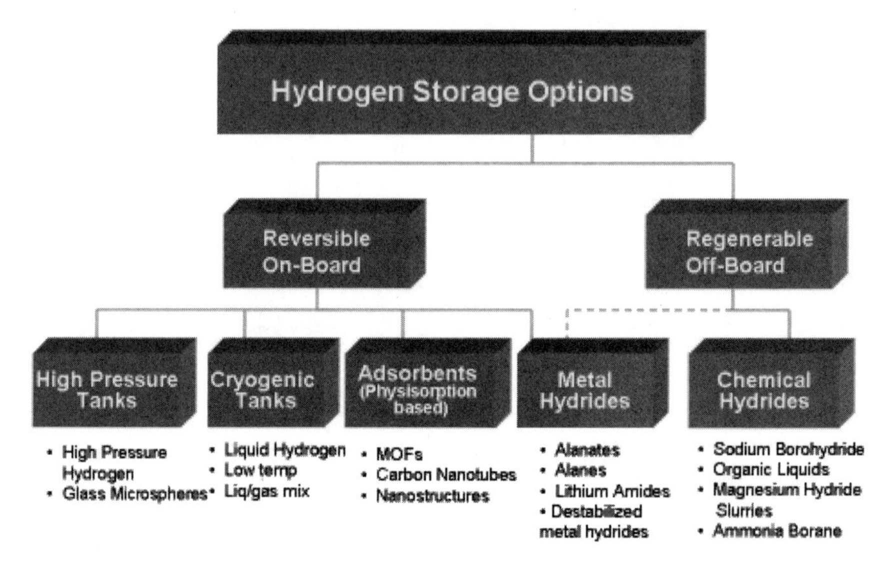

FIGURE 4 Options for vehicular hydrogen storage. Source: U.S. DOE.

High-pressure and cryogenic tanks meet some of the near-term DOE targets and are already in use in prototype vehicles. The state of the art in high-pressure tanks is 10,000 psi (or about 700 atm), as developed by Quantum and others (Ko and Newell, 2004). Remaining challenges include volumetric capacity and issues related to high pressure and cost. Refueling times, compression energy penalties, and heat-management during compression must also be addressed because the mass and pressure of onboard hydrogen would have to be increased to provide a driving range of more than 300 miles.

Cryogenic, or liquid, hydrogen (LH_2) tanks can, in principle, store more hydrogen in a given volume than compressed tanks, because the volumetric capacity of liquid hydrogen is about 0.07 kg/L (compared to roughly 0.04 kg/L even at 700 atm). Key issues related to LH_2 tanks are hydrogen boil-off, the energy required for hydrogen liquefaction (typically 35 percent of the lower heating value of hydrogen), insulation requirements, and cost.

Metal hydrides function by dissociatively adsorbing (or absorbing) hydrogen into their metal lattices, thereby allowing for higher energy densities than liquid hydrogen. Figure 5 shows that the optimum "operating P-T window" for PEM fuel cell vehicular applications is in the range of 1–10 atm and 25–120°C (Wang et al., 2004). A simple metal hydride, such as $LaNi_5H_6$, that incorporates hydrogen into its crystal structure, can function in this range, but its gravimetric capacity is too low and its cost is too high for vehicular hydrogen storage applications. However, at the present time, $LaNi_5H_6$ is one of the few commercially available metal hydrides.

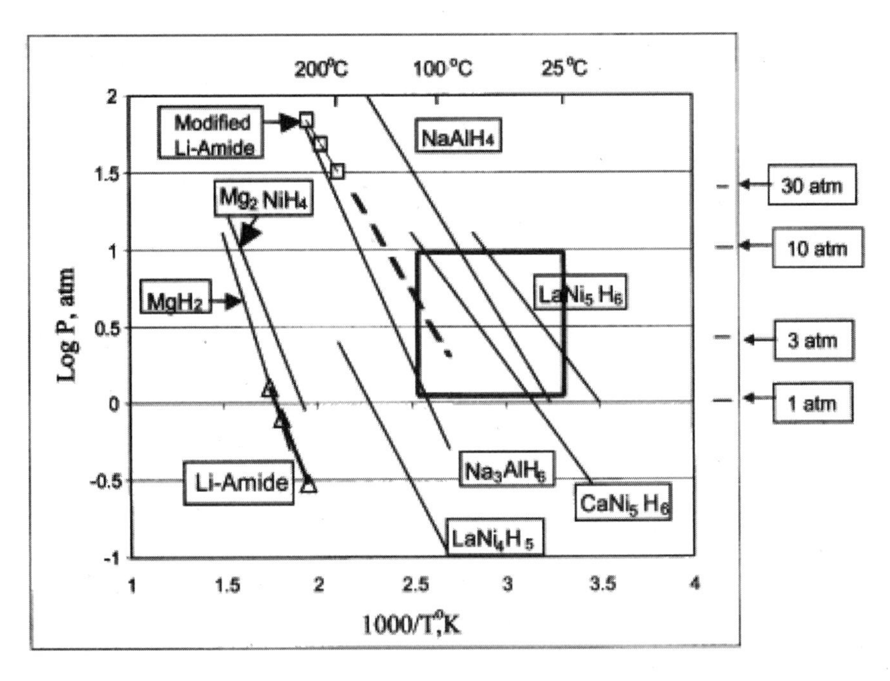

FIGURE 5 Equilibrium pressure-temperature curves for various metal hydride materials. The P-T window for PEM fuel cell vehicular applications is indicated by the highlighted box. Source: Wang et al., 2004.

Metal hydrides have been studied for decades, and Sandrock and Thomas (2001) have compiled a database of metal hydride properties to help guide the development of improved materials (see also SNL, 2005). Because most metals in traditional metal hydrides are heavy, the gravimetric capacity of such systems is unacceptable. However, in 1997, a breakthrough by Bogdanovic and Schwickardi (1997) demonstrated that titanium species could act as a catalyst in reversibly storing hydrogen in "complex" metal hydrides (e.g., $NaAlH_4$). Such systems, with light elements (in this case Na and Al), can achieve much higher gravimetric capacities without compromising volumetric capacities. Although $NaAlH_4$ cannot meet the DOE targets, this recent discovery has spurred activity around the world on complex metal hydrides.

New systems, such as $Li_2NH + H_2 = LiNH_2 + LiH$, have recently been discovered (Chen et al., 2002; Luo, 2004; Nakamori and Orimo, 2004). The substitution by light metals, such as Mg, is an active area of research to adjust operating temperatures and pressures and improve kinetics. Another promising discovery in 2005 by Vajo and coworkers is that metal hydrides such as $LiBH_4$ can be "destabilized" to achieve high capacity at lower pressures and temperatures than previously known. Vajo's $LiBH_4/MgH_2$ material showed a hydrogen storage capacity of 10 weight percent, which is prompting attention worldwide on this approach (Vajo et al., 2005).

One of the main issues related to metal hydrides is that the heat of reaction is typically 30–40 kJ/mol. This means that, to meet refueling time targets (~3 minutes for 5 kg H_2), close to 0.5 MW must be rejected during the charging of typical metal hydride systems. Thus, certain metal hydrides, such as AlH_3, appear to be more suitable for off-board regeneration. However, significant engineering challenges lie ahead, such as thermal management and reactor design optimization to meet weight, volume, and cost targets. Other issues related to metal hydrides include low gravimetric capacity and slow uptake and release kinetics.

Chemical hydrogen storage refers to chemical reactions, such as the hydrolysis of sodium borohydride ($NaBH_4 + 2H_2O \rightarrow NaBO_2 + 4H_2$), which has been demonstrated by Millennium Cell and others (Amendola et al., 2000; Wu et al., 2004), or the dehydrogenation of organic compounds, such as methylcyclohexane or decalin, which have been studied for decades, particularly in Japan. In 2004, a new type of liquid-phase hydrogen storage material was demonstrated. This liquid, based on N-ethylcarbazole, can attain more than 5.5 weight percent and 0.05 kg/L of hydrogen storage; several dehydrogenation/ hydrogenation cycles of N-ethylcarbazole have recently been shown (Figure 6)

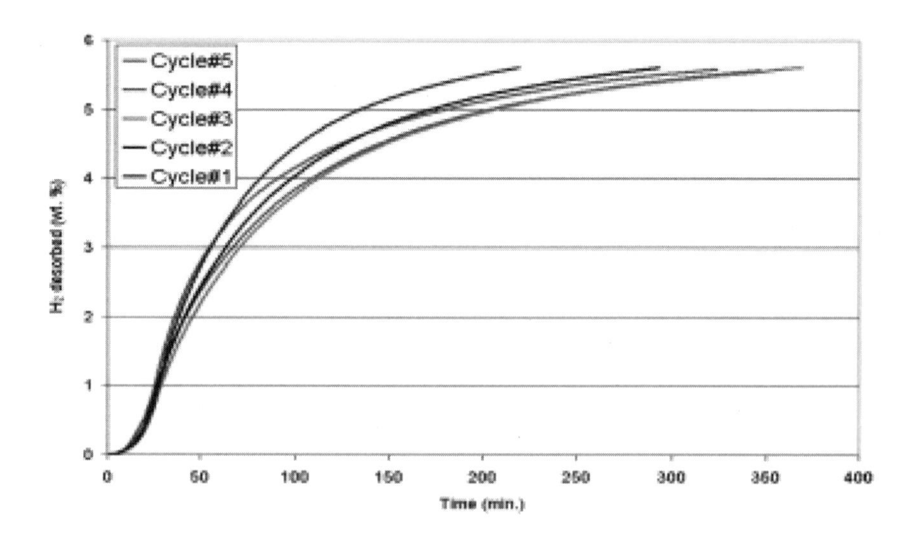

FIGURE 6 Several cycles demonstrated for the dehydrogenation of N-ethylcarbazole. Source: Cooper et al., 2004.

(Cooper et al., 2004). Another exciting recent discovery is that, by combining ammonia borane (NH_3BH_3) in a mesoporous scaffold structure, the hydrogen-release reaction can be tailored in terms of by-product release and temperature; roughly 6 weight percent hydrogen storage has been demonstrated (Gutowska et al., 2005).

The most significant issue related to chemical hydrogen storage to be addressed is that the covalent bonds broken to release hydrogen cannot be easily replaced on board a vehicle. The "spent" fuel must be reclaimed from the car and regenerated off board at a central plant or at the fueling station. The energy requirements and total life-cycle analysis, including environmental impact, are under study.

Finally, high-surface-area sorbents, such as nanostructured carbon materials or metal organic frameworks (MOFs), and perhaps clathrates, can be used to adsorb hydrogen physically (Dillon et al., 2004; NREL, 2005). There is still some controversy as to how much hydrogen some of these materials can store. Although they have potentially high gravimetric capacity, because they have high surface areas (e.g., ~4,000 m^2/g), their volumetric capacity for hydrogen storage will probably be low. However, one advantage of these materials is that the binding of hydrogen is weak so the release of hydrogen should not require high temperatures. Therefore, power consumption during hydrogen release, as well as heat rejection during refueling, would not be as challenging as for metal hydrides. Engineering issues, such as reactor design, as well as tuning fundamental material properties must also be addressed.

With improved theoretical modeling, combinatorial/high-throughput screening techniques and understanding at the nanoscale, future work worldwide will focus on "tailoring" materials to meet the targets for hydrogen storage. In addition to activities in the United States, DOE is supporting the mission of the International Partnership for the Hydrogen Economy (IPHE), formed in November 2003, to help accelerate global activities to achieve a hydrogen economy. Working with other countries, DOE has helped organize IPHE conferences on hydrogen storage, hydrogen production, and other research areas to identify, evaluate, and coordinate multinational R&D and demonstration programs to accelerate the advancement of hydrogen and fuel cell technologies (IPHE, 2005).

REFERENCES

Amendola, S.C., S.L. Sharp-Goldman, M.S. Janjua, M.T. Kelly, P.J. Petillo and M. Binder. 2000. An ultrasafe hydrogen generator: aqueous, alkaline borohydride solutions and Ru catalyst. Journal of Power Sources 85(2): 186–189.

Bogdanovic, B., and M. Schwickardi. 1997. Ti-doped alkali metal aluminium hydrides as potential novel reversible hydrogen storage materials. Journal of Alloys and Composites 253–254: 1–9.

Chen, P., Z. Xiong, J. Luo, J. Lin, and K.L. Tan. 2002. Interaction of hydrogen with metal nitrides and imides. Nature 420(6913): 302–304.

Cooper, A., G. Pez, H. Cheng, A. Scott, D. Fowler, and A. Abdourazak. 2004. Design and Development of New Carbon-based Sorbent Systems for an Effective Containment of Hydrogen. Pp. 210–214 in DOE Hydrogen Program FY 2004 Progress Report. Washington, D.C.: U.S. Department of Energy. Available online at: *http://www.eere.energy.gov/hydrogenandfuelcells/pdfs/annual04/iiib4_pez.pdf.*

Devlin, P. 2005. 2005 Annual DOE Hydrogen Program Review: Hydrogen Production and Delivery. PowerPoint presentation. Available online at: *http://www.hydrogen.energy.gov/pdfs/review05/pd1_devlin.pdf.*

Dillon, A.C., P. Parilla, T. Gennett, K. Gilbert, J. Blackburn, Y.-H. Kim, Y. Zhao, S. Zhang, J. Alleman, K. Jones, T. McDonald, and M. Heben. 2004. Hydrogen Storage in Carbon-Based Materials. Pp. 245–252 in DOE Hydrogen Program FY 2004 Progress Report. Washington, D.C.: U.S. Department of Energy. Available online at: *http://www.eere.energy.gov/hydrogenandfuelcells/annual_report04_storage.html#carbon.*

DOE. 2005a. Annual Energy Outlook 2005 with Projections to 2025. Available online at: *http://www.eia.doe.gov/oiaf/aeo/index.html.*

DOE. 2005b. Hydrogen Posture Plan: An Integrated Research, Development, and Demonstration Plan. Available online at: *http://www.eere.energy.gov/hydrogenandfuelcells/pdfs/hydrogen_posture_plan.pdf.*

DOE. 2005c. Multi-Year Research, Development, and Demonstration Plan: Planned Program Activities for 2003–2010. Hydrogen, Fuel Cells and Infrastructure Technologies Program. Available online at: *www.eere.energy.gov/hydrogenandfuelcells/mypp.*

DOE. 2005d. Hydrogen Storage: Current Technology. Hydrogen, Fuel Cells and Infrastructure Technologies Program. Available online at: *http://www.eere.energy.gov/hydrogenandfuelcells/storage/current_technology.html.*

Gutowska, A., L. Li, Y. Shin, C.M. Wang, X.S. Li, J.C. Linehan, R.S. Smith, B.D. Kay, B. Schmid, W. Shaw, M. Gutowski, and T. Autrey. 2005. Nanoscaffold mediates hydrogen release and the reactivity of ammonia borane. Angewandte Chemie International Edition 44(23): 3578–3582.

IPHE (International Partnership for the Hydrogen Economy). 2005. IPHE Updates. Available online at: *under www.iphe.net.*

Ko, J., and K. Newell. 2004. Low-Cost, High-Efficiency, High-Pressure Hydrogen Storage. Pp. 183–185 in DOE Hydrogen Program FY 2004 Progress Report. Washington, D.C.: U.S. Department of Energy. Available online at: *http://www.eere.energy.gov/hydrogenandfuelcells/pdfs/annual04/iiia1_ko.pdf.*

Luo, W. 2004. ($LiNH_2$–MgH_2): a viable hydrogen storage system. Journal of Alloys and Compounds 381(1-2): 284–287.

Nakamori, Y., and S. Orimo. 2004. Destabilization of Li-based complex hydrides. Journal of Alloys and Compounds 370(1-2): 271–275.

NREL (National Renewable Energy Laboratory). 2005. DOE National Center for Carbon-Based Hydrogen Storage. Available online at: *http://www.nrel.gov/basic_sciences/carbon_based_hydrogen_center.html.*

Ordaz, G., J. Petrovic, C. Read, and S. Satyapal. 2005. 2005 Annual DOE Hydrogen Program Review: Hydrogen Storage. PowerPoint presentation. Available online at: *http://www.hydrogen.energy.gov/pdfs/review05/ st1_satyapal.pdf.*

Sandrock, G., and G. Thomas. 2001. The IEA/DOE/SNL on-line hydride databases. Applied Physics A 72(2): 153–155.

SNL (Sandia National Laboratories). 2005. Hydride Information Center. Available online at: *http://hydpark.ca.sandia.gov/.*

Vajo, J. , S. Skeith, and F. Mertens. 2005. Reversible Storage of Hydrogen in Destabilized $LiBH_4$. Journal of Physical Chemistry B 109: 3719-3722.

Wang, J., M. Allendorf, R. Baldonado, B. Bastesz, T.Boyle, S. Daniel, D. Dedrick, K. Gross, S. Karim, J. Liu, W. Luo, E. Majzoub, T. McDaniel, T. Nenoff , M. Phillips, J. Reilly, G. Sandrock, S. Spangler, R. Stumpf, K. Thürmer, and J. Voigt. 2004. Hydride Development for Hydrogen Storage. Pp. 220–229 in DOE Hydrogen Program FY 2004 Progress Report. Washington, D.C.: U.S. Department of Energy. Available online at: *http://www.eere.energy.gov/hydrogenandfuelcells/pdfs/annual04/iiic2_wang.pdf.*

Wu, Y., M.T. Kelly, and J.V. Ortega. 2004. Low-Cost, Off-Board Regeneration of Sodium Borohydride. Pp. 195–199 in DOE Hydrogen Program FY 2004 Progress Report. Washington, D.C.: U.S. Department of Energy. Available online at: *http://www.eere.energy.gov/hydrogenandfuelcells/pdfs/annual04/iiib1_wu.pdf.*

Fuel Cells: Current Status and Future Challenges

STUART B. ADLER
University of Washington
Seattle, Washington

Fuel cells, which convert chemical energy directly to electricity, are more efficient than current means of energy conversion. The question is where they might fit in the broad spectrum of energy choices. This paper briefly reviews and compares polymer-electrolyte fuels cells (PEFCs) and solid-oxide fuel cells (SOFCs) and then describes significant scientific challenges that must be overcome before these technologies can become commercially competitive.

Fuel cells are not a new idea. Sir William Grove first demonstrated the conversion of hydrogen to electricity using an acid-electrolyte fuel cell in 1839. However, turning this idea into a practical means of energy conversion has proved to be elusive. A major technical and cost barrier has been implementation of liquid electrolytes, the basis for most commercial fuel cells (e.g., alkaline fuel cells, molten-carbonate fuel cells). In contrast, the fuel cells of greatest commercial interest today are based on solid electrolytes, which have benefited from recent advances in materials and manufacturing.

For the purposes of discussion, we can divide solid-electrolyte fuel cells into two types: (1) PEFCs, often referred to as proton-exchange-membrane (or PEM) fuel cells; and (2) SOFCs. Figure 1 illustrates how these types of fuel cells function.

A common justification for fuel cells has been environmental protection—the idea that fuel cells produce only water as a combustion by-product and thus are "zero emission" devices. However, it is difficult to make the case for fuel cells based on this argument alone. Although fuel cells themselves produce only

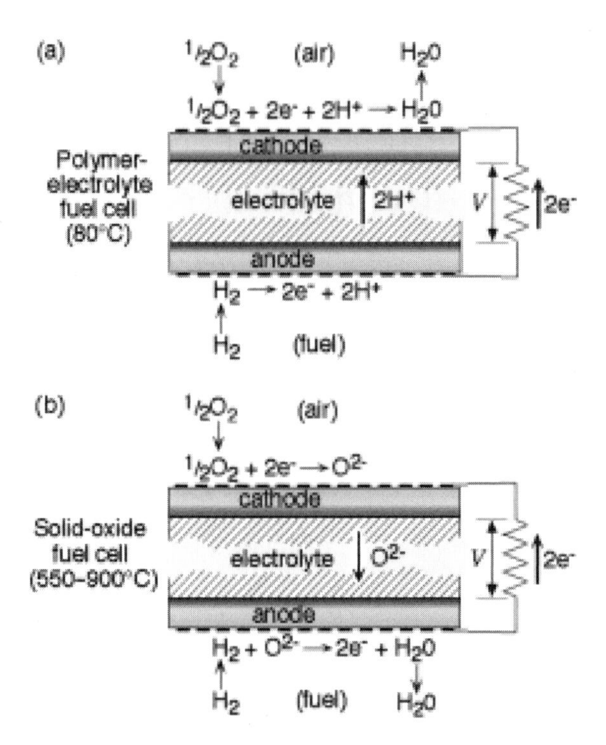

FIGURE 1 Two types of solid electrolyte fuel cells. a. In a PEFC, a proton-conducting polymer membrane is exposed on one side to fuel (hydrogen) and on the other to air. On the hydrogen side (anode), H_2 gas is oxidized, and the protons thus created migrate to the other side of the membrane (cathode), where O_2 gas in the air is reduced to water. Some portion of the reversible work of the net reaction is recovered as a voltage difference between cathode and anode, delivered to an external circuit by the flow of electrons. PEFCs typically operate at 80~100°C. b. In an SOFC, a ceramic oxygen ion conductor at elevated temperatures (500~1,000°C) serves as the electrolyte membrane. In this case, the fuel (which can be a mixture of H_2, CO, and/or hydrocarbons) is oxidized to H_2O and CO_2 at the anode, while O_2 is reduced to O^{2-} at the cathode. In both types of fuel cells, cells are normally assembled into multicell stacks to increase system voltage and provide a means of distributing gases (fuel and air) evenly.

water, the production of hydrogen from hydrocarbons, such as oil or coal, involves the production of carbon dioxide (CO_2) and requires the suppression of sulfur dioxide (SO_2). Thus fuel cells merely transfer the environmental problem elsewhere.

In addition, numerous technologies are already available that can eliminate SO_2 and nitrogen oxides (NO_X) from combustion. Widespread implementation of these technologies is simply a matter of cost and political will. Thus, one can easily imagine an energy economy based entirely on clean combustion of hydrogen or other multisource fuels that does not include fuel cells.

To understand the potential role of fuel cells, we must instead consider their primary advantage—efficiency. In this regard, fuel cells are an enabling (rather than a displacing) technology. They recover energy that is normally lost by the irreversible process of combustion. Thus, fuel cells offer a potential path toward overall reduction of fuel consumption that combustion simply cannot provide, even after many years of incremental improvements.

By reducing the overall amount of CO_2 produced per kilowatt (kW) of usable power, increased efficiency may, ultimately, have environmental benefits as well. In addition, the required retooling of the fuel infrastructure toward more generic, small-molecule fuels (e.g., H_2, CO, CH_4) might also lead to centralization of CO_2 production, which would facilitate carbon sequestration and reduce the vulnerability of particular energy sectors to fluctuations in the supply of particular fuel sources (e.g., the dependence of gasoline prices on the availability of oil from the Middle East).

COMPARISONS BETWEEN PEFCs AND SOFCs

A primary factor influencing the trade-off between capital and efficiency in fuel cells is operating temperature. SOFC stacks, which operate at temperatures ranging from 550°C to 900°C, produce high-quality waste heat that can be captured for increased efficiency, combined heat and power, or reformation of hydrocarbons (HCs). SOFC stacks tend to operate adiabatically wherein excess air is used as the primary coolant, and thus heat can be recovered from the SOFC exhaust. This feature has made SOFCs very attractive for the production of stationary power, where efficiency is of high importance relative to capital cost, and operation on reformed HCs is an advantage. Allowable capital costs for stationary power ($400/kW) are about 10 times higher than for PEFCs in automotive applications (DOE, 2004b).

By using thin-film ceramics supported on low-cost metal alloys, SOFC developers have reduced material and manufacturing costs, lowered operating temperatures, and significantly mitigated cell-degradation problems. Figure 2 shows an example of a metal-supported cell based on a thin ceria electrolyte, capable of stable power densities of ~500 mW/cm2 at 570°C (Brandon, 2005). Systems based on this type of cell are nearing efficiency and cost targets for use in homes (combined heat and power) and auxiliary power units for trucks and aircraft.

In contrast, PEFCs have historically been designed to operate isothermally, at or below 80°C. Low operating temperatures have made them more suitable for small or mobile applications, for which capital cost requirements are much more stringent, pure hydrogen (H_2) is assumed to be available, and the efficiencies of heat integration are less important. The most challenging market from a capital-cost perspective is motive power (cars), for which allowable capital costs are

FIGURE 2 Example of a metal-supported, thin-film solid-oxide fuel cell capable of operation below 600°C. Photo courtesy of Ceres Power, Ltd., reproduced with permission.

estimated to be on the order of $35/kW (Garman, 2003). PEFCs are also generally thought to match the size, weight, and start-up constraints for primary power in automobiles.

Substantial progress has been made in increasing the power density of PEFCs (>1 kW/kg) (Gasteiger et al., 2005), as well as reducing the amount of platinum (Pt) catalyst to a level that is reasonable to recycle (<15g/vehicle, three to four times the catalyst in a catalytic converter) (Cooper, 2004; Gasteiger et al., 2005). Based on these successes, several of the world's biggest automakers, including General Motors, Ford, Daimler, and Honda (Figure 3), have built demonstration cars.

Despite these significant advances, solid-electrolyte fuel cells have not yet achieved widespread penetration into the energy market for many reasons. In particular, fuel cell systems are still too costly to be competitive with existing technology at current energy prices. Although this situation may change as fuel prices rise and capital costs come down with manufacturing improvements and economies of scale, fundamental technological barriers must also be overcome before cost reductions are likely. Many of these technological hurdles have been described in detail elsewhere (DOE, 2004a). The discussion below focuses on areas of fundamental research where breakthroughs might lead to significant technological advancements.

FIGURE 3 Sandy Spallino, first individual customer to purchase a PEFC-powered car, fills up her Honda FCX at one of many H_2 refueling stations planned for California. Source: Honda press release, June 2005.

MATERIAL PROPERTIES BY DESIGN

Many of the materials used in SOFCs and PEFCs today are similar to the ones used 25 years ago. Examples include the nickel (Ni)-cermet anode used in most SOFCs and the perfluorosulfonic acid (PFSA) membrane used as the electrolyte in most PEFCs (Dupont Nafion®). Despite numerous difficulties with these materials, they are still considered state of the art because their unique combination of properties is still unmatched. However, they also introduce fundamental problems (Figure 4). In SOFCs, Ni-cermet has very poor sulfur tolerance, especially below 800°C, which makes it unsuitable as a long-term SOFC anode (DOE, 2004a). PEFC developers have concluded, that to be successful in cars, the system must operate at 110~120°C, which introduces severe performance and degradation problems for PFSA (Gasteiger and Mathias, 2003). To date, a trial and error approach has been used to search for new materials. However, further advances are likely to require a directed design approach (Hickner et al., 2004) and/or combinatorial methods (Kilner et al., 2005).

PROBING AND CONTROLLING MICROSTRUCTURE/NANOSTRUCTURE

Despite the technological advances in SOFC and PEFC technology in the last ten years, our understanding and design capability are mostly at the macro-

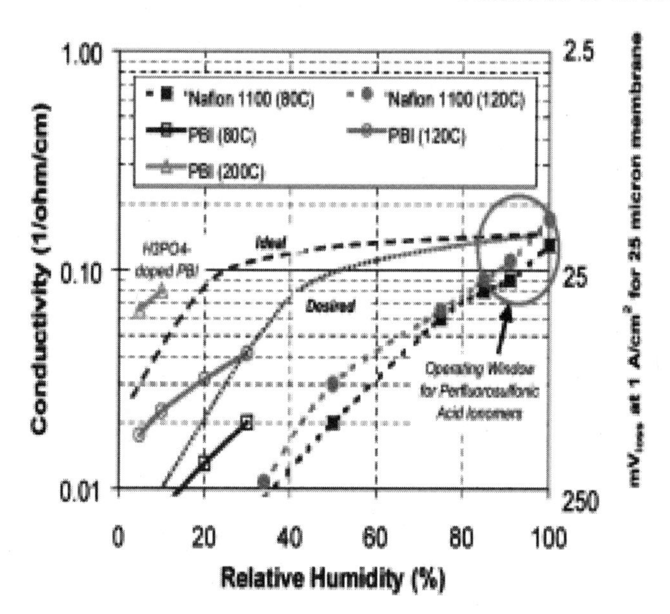

FIGURE 4 Relationship between proton conductivity and relative humidity in the adjoining gas at various temperatures for PFSA and phosphoric-acid-doped polybenzimidazole (PBI). Curves are also shown for materials that would enable and be ideal for system simplification. Source: Gasteiger and Mathias, 2003.

scopic/empirical level. The microstructure of a PEFC electrode, for example, is still understood only in a very general sense; exactly how the catalyst, ionomer, and gas come together and affect performance is generally not well understood and thus not amenable to intelligent design. For example, one proposed strategy for improving the catalyst in PEFC cathodes is to concentrate Pt particles near the opening of the aqueous flow channel in the PFSA ionomer; at present they are distributed randomly throughout the electrode matrix. However, this type of nanostructural analysis, let alone control, is not possible today.

As shown in Figure 5, one possible technique on the horizon for SOFCs is focused-ion beam milling coupled with electron microscopy or other surface analytical techniques, which may make it possible to analyze and direct electrode microstructures in new ways (J. Wilson et al., 2005). Researchers have also recently demonstrated solution impregnation of materials into an electrolyte host matrix to obtain SOFC electrodes with improved hydrocarbon activity or O_2 reduction performance (Huang et al., 2005; McIntosh and Gorte, 2004).

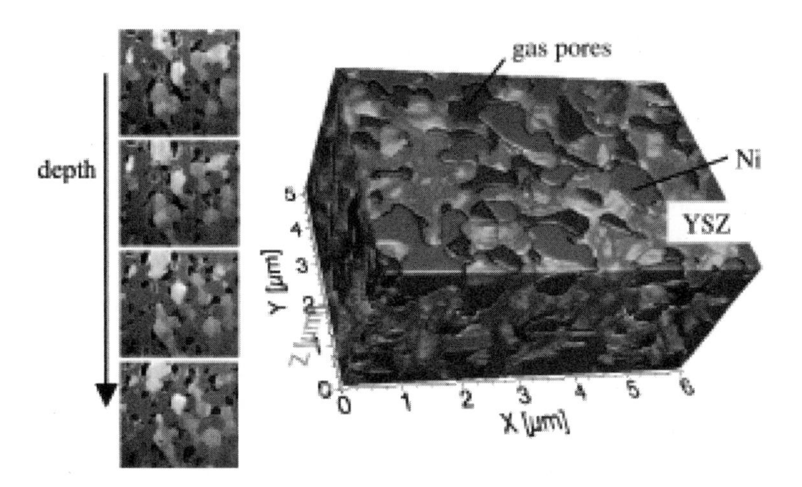

FIGURE 5 3-D reconstruction of pores inside a porous Ni-YSZ SOFC anode, based on 2-D FIB-SEM image slices. At left are cross-sectional image slices corresponding to the reconstructed 3-D image.

UNDERSTANDING ELECTRODE DEGRADATION AND OTHER DEGRADATION PROCESSES

The vast majority of work in the last 10 years has been focused on improving fuel cell performance. However, as the technology has now reached some performance targets, and as more cells and stacks have been tested for longer periods of time, long-term durability has risen to the top of the list of performance targets. For example, SOFC electrodes can be very sensitive to chromia (Cr) poisoning (Simner and Stevenson, 2004). Although electrode degradation has been positively linked to Cr contamination from metal interconnects, it is not clear why some electrode materials are more sensitive than others or why seemingly similar electrodes tested by different groups degrade at different rates. The answers to these questions require a much deeper mechanistic and scientific understanding of electrode processes than we currently possess.

Recent advances in microfabrication and diagnostics may significantly improve our ability to control and analyze electrode reactions (Adler, 2004; J.R. Wilson et al., in press). Recent work using nonlinear electrochemical impedance spectroscopy to resolve SOFC cathode reaction mechanisms may eventually improve our ability to diagnose how and why electrodes degrade and guide the selection of new materials and fabrications to mitigate degradation.

OUTLOOK

Fuel cells continue to face major technological hurdles that may require many years of research and development before they can be overcome. In addition, fuel cells are not likely to be implemented in isolation. They must be part of a larger shift in fuel infrastructure and efficiency standards, which will require sustained political and economic pressure—and time. Finally, like any technology, economy of scale will require a natural maturation process over many years or decades (DeCicco, 2001).

Taken together, these hurdles suggest that the widespread adoption of fuel cell technology is not likely in the short term. Successful advancement of fuel cell technology will require a sustained, long-term commitment to fundamental research, commercial development, and incremental market entry.

REFERENCES

Adler, S.B. 2004. Factors governing oxygen reduction in solid oxide fuel cell cathodes. Chemical Reviews 104(10): 4791–4844.

Brandon, N.P. 2005. Metal Supported Solid Oxide Fuel Cells for Operation at 500–600°C. Paper presented at SSI-15, the International Conference on Solid State Ionics, Baden-Baden, Germany, July 17–22, 2005.

Cooper, J.S. 2004. Recyclability of Fuel Cell Power Trains. Pp. 792–801 in Proceedings of the SAE 2004 World Congress. SAE Technical Paper 2004-01-1136. Warrendale, Pa.: SAE International.

DeCicco, J. 2001. Fuel Cell Commercialization Perspectives: Market Context and Competing Technologies. Paper presented at the 2001 Gordon Conference on Fuel Cells, Bristol, R.I., July 29–August 3, 2001.

DOE (U.S. Department of Energy). 2004a. Fuel Cell Handbook, 7th ed. National Energy Technology Laboratory, Morgantown, West Virginia. Washington, D.C.: Office of Fossil Energy, U.S. Department of Energy.

DOE. 2004b. Fuel Cell Program Annual Report. National Energy Technology Laboratory, Morgantown, West Virginia. Washington, D.C.: Office of Fossil Energy, U.S. Department of Energy.

Garman, D. 2003. Testimony by David K. Garmen, Assistant Secretary, Energy Efficiency and Renewable Energy, U.S. Department of Commerce. Hearing before the Committee on Science, 108th Congress, Serial Number 108-4, March 5, 2003. Available online at: *http://commdocs.house.gov/committees/science/hsy85417.000/hsy85417_0f.htm.*

Gasteiger, H.A., and M.F. Mathias. 2003. Fundamental Research and Development Challenges in Polymer Electrolyte Fuel Cell Technology. Paper presented at the Workshop on High Temperature PEM Fuel Cells, Pennsylvania State University, December 2003.

Gasteiger, H.A., S.S. Kocha, B. Sompalli, and F.T. Wagner. 2005. Activity benchmarks and requirements for Pt, Pt-alloy, and non-Pt oxygen reduction catalysts for PEMFCs. Applied Catalysis B: Environmental 56(1–2): 9–35.

Hickner, M.A., H. Ghassemi, Y.S. Kim, B.R. Einsla, and J.E. McGrath. 2004. Alternative polymer systems for proton exchange membranes (PEMs). Chemical Reviews 104(10): 4637–4678.

Huang, Y.Y., J.M. Vohs, and R.J. Gorte. 2005. Characterization of LSM-YSZ composites prepared by impregnation methods. Journal of the Electrochemical Society 152(7): A1347–A1353.

Kilner, J.A., J.C.H. Rossiny, and S. Fearn. 2005. Combinatorial Searching for Novel Mixed Conductors. Paper presented at SSI-15 International Conference on Solid State Ionics, Baden-Baden, Germany, July 17–22, 2005.

McIntosh, S., and R.J. Gorte. 2004. Direct hydrocarbon solid oxide fuel cells. Chemical Reviews 104(10): 4845–4865.

Simner, S.P., and J.W. Stevenson. 2004. Cathode-Chromia Interactions. Paper presented as part of the SECA 2004 Annual Meeting and Core Program Review, Boston, Massachusetts 2004.

Wilson, J., W. Kobsiriphat, R. Mendoza, J. Hiller, D. Miller, K. Thornton, P. Voorhees, S. Adler, and S. Barnett. 2005. Three Dimensional Reconstruction of Solid Oxide Fuel Cell Electrodes. Unpublished paper.

Wilson, J.R., D.T. Schwartz, and S.B. Adler. In press. Nonlinear electrochemical impedance spectroscopy for mixed-conducting SOFC cathodes. Electrochemica Acta. Available online at: *http://dx.doi.org/doi:10.1016/j.electacta.2005.02.109.*

Dinner Speech

Engineering for a New World

SHIRLEY ANN JACKSON
Rensselaer Polytechnic Institute
Troy, New York

Good evening. It is a pleasure and an honor to be here at this illustrious and select gathering of the nation's top engineers. I am engaged by the knowledge that you—chosen because you are the best, the brightest, and at an age when you have accomplishments to your credit and productive years ahead—are here to be challenged.

I have always noted that it is risky to make predictions about the future, especially on a global scale. The impact of key events of the past five years—the terrorists attacks of September 2001 in the United States, the SARS epidemic, and the recent earthquake and tsunami—tell us that the twenty-first century may turn out much differently than our best prophets could have predicted.

One key challenge of the twenty-first century rose up and looked us directly in the eyes a few weeks ago when Hurricane Katrina devastated parts of the U.S. Gulf Coast. And now Hurricane Rita is threatening to do the same. As we know, Hurricane Katrina swamped cities; cut power lines; closed shipping ports; damaged oil drilling and refining facilities, knocking out about 10 percent of U.S. refining capacity (Lucchetti et al., 2005); and left hundreds dead and hundreds of thousands homeless. The mass relocation of people, many have said, is the largest since the Dust Bowl out-migration of the 1930s or the dislocation caused by the Civil War.

The disruption of key energy systems in the Gulf region rippled throughout the nation and the economy. The oil industry strained to recover oil rigs and refineries. There were some gas stations without gasoline, long lines at others, and gasoline prices soared into uncharted territory. The U.S. Postal Service turned away mail addressed to zip codes in the affected areas. Wire services advised against e-mailing to southern Louisiana, Mississippi, and parts of Alabama because of "bounce-back" volume. The educations of 75,000 college students in the region were interrupted. The effects are still being revealed, understood, evaluated, and comprehended.

Among many lessons—some that will be long discussed and debated—the storm brought home to the U.S. population, perhaps in a new way, the costs, both economic and social, of a major disruption of our basic infrastructure. As oil and gas companies and utilities struggle to get "back on line," Hurricane Katrina clearly illustrates our dependence on a readily available, inexpensive, uninterrupted supply of energy.

Yet, the impact of Hurricane Katrina on U.S. energy supplies only made clearer a global situation that is steadily building—namely, the critical need for energy security, not only in this nation, but, indeed, throughout the entire world.

GLOBAL ENERGY OUTLOOK

Although a looming global energy-security crisis was laid bare by disaster, it is being accelerated by a positive force—extraordinary economic growth in many developing nations. This growth is enabling them to provide their populations with the common necessities of life—food, shelter, clothing, transportation, and education—necessities to which many never had access before. This unprecedented growth is both enabled by energy availability, and, at the same time, causing a heretofore unparalleled demand for energy in all of its forms. Global energy consumption is projected to increase by 57 percent from 2002 to 2025 (EIA, 2005). But another point to consider is that for every two gallons of petroleum-based fuel consumed, one gallon is discovered.

In the past 35 to 40 years, worldwide energy consumption has nearly doubled, driven by population growth, rising living standards, the invention of energy-dependent technologies, and consumerism. Energy consumption has increased nearly everywhere, with the most dramatic percentage increase in China and the rest of Asia. Coal usage has decreased marginally, but consumption of every other major energy source has increased markedly. Electricity use has nearly tripled. If these trends continue, global energy consumption will double again by mid-century. Fossil fuels will continue to dominate, and the share of nuclear power and renewable energy sources—wind, solar, and geothermal energy—will remain limited (IEA, 2004).

Although the planet has enough energy resources to meet this demand beyond 2030 (IEA, 2004), it is less certain how much it will cost to extract and deliver these fuels to consumers. New energy infrastructure will require vast amounts of financing. Fossil fuels are projected to account for about 85 percent of the increase in consumption. Major oil and gas importers—including the

United States, Western Europe, and the expanding economies of China and India—will become more dependent on supplies from Middle East members of OPEC and Russia. As international trade expands, the vulnerability to disruptions will increase, and geopolitical turmoil may exacerbate surging energy prices. In addition, carbon dioxide (CO_2) emissions will continue to rise, calling into question the sustainability of current energy usage models.

An estimated 1.6 billion people do not have access to electricity (IEA, 2004). One-sixth of the world's population does not have safe drinking water (Osborne, 2005); one-half do not have adequate sanitation; and one-half live on less than $2 per day. A reliable energy supply—especially electricity is a prerequisite for addressing these needs—and is the basis of the United Nations Millennium Goals that were set five years ago.

In many developing countries, the energy-poverty levels are severe. China, however, is a success story in the making. Throughout the 1990s, Chinese electricity generation grew at an average rate of 8 percent per year. In 2003, electricity generation in China increased by 16 percent, and in 2004 the rate of increase was even higher (~18 percent) (IEA, 2004).

The increase in oil consumption in China, from 2002 to 2003, accounted for more than 18 percent of the increase in global oil demand—and, in the process, China surpassed Japan and became the second largest oil consumer. In fact, China is the second largest consumer of primary energy overall—not to mention the second largest economy and the second largest contributor to energy-related CO_2 emissions. If projections hold, China will continue to dominate growth in energy demand. It should come as no surprise, then, that the 10th Five-Year Plan of the Chinese government, covering the period 2001 to 2005, puts energy conservation near the top of the energy policy agenda. In fact, China is about to introduce more stringent fuel-economy standards for new vehicles than those in force in the United States (IEA, 2004).

The real paradox is that something to be celebrated—the continuing progress of developing nations and the human progress it represents—has yet to command the attention of the global community, which must deal with the issues of global resources and energy distribution and the need for alternatives to fossil fuels. Attempts so far have focused only on certain aspects of the problem. The new solutions, however, will require a holistic approach.

This is just a thumbnail sketch of a vast, complex, and interconnected issue. It is not my intention to discuss the complexities and exigencies of the near-term energy crisis caused by Hurricane Katrina. Rather, I will use the energy squeeze Katrina caused as an example that should compel us, as engineers and scientists, to step up to the challenge. And, in so doing, we must strive to be truly international—to think in a global way. We have a moral and social responsibility to address this issue, not just for a single nation or for a temporary fix, but to solve one of humanity's most urgent challenges. The science and engineering communities should be well positioned to do this because they have been global communities for hundreds of years.

Energy security underlies all progress, because, of course, virtually all technological advances in the past 150 years have been predicated upon readily available energy sources and technologies. Energy security is, then, a key global challenge—one that will require global perspectives, global thinking, global solutions, and innovation of the highest order. Indeed, this same perspective and approach must prevail as we seek solutions to other global "threats without borders," including infectious diseases—such as SARS, AIDS, and avian flu. Like Hurricane Katrina, the threats include natural disasters, such as last December's tsunami in Southeast Asia. They include global climate change, species extinction, and acid rain, among others. They include terrorism and the myriad challenges facing a significant segment of the global population that does not have the basic needs of life—sufficient food, clean water, health care, and education.

The global nature of these challenges provides a measure of the urgency of advancing discovery and innovation to resolve them. It is a given that, at least in the long term, no single "solution" will ensure abundant, clean, and inexpensive energy for the global community. There is likely to be a "mix" of solutions, including innovative extractive and transportation technologies for fossil fuels, innovative conservation technologies, and innovative alternative-fuel technologies.

I will not attempt to review the full spectrum of energy technologies currently under consideration, some of which you are discussing at this symposium, but I will examine a few—nuclear power, hydrogen and fuel cells, and fusion.

NUCLEAR POWER

I will start with an old/new technology—nuclear power, which currently generates 16 percent of global electricity—about 20 percent in the United States; 17 percent in Russia; 3.3 percent in India; 2.2 percent in China; and about 30 percent in Western Europe (provided by about 150 nuclear power plants) (ElBaradei, 2005). Nuclear power produces virtually no sulfur dioxide, particulates, nitrogen oxides, volatile organic compounds, or greenhouse gases. The complete cycle, from resource extraction to waste disposal—including facility and reactor construction—emits only 2 to 6 grams of carbon equivalent per kilowatt-hour. This is about the same as wind and solar power, if we include construction and component manufacturing. All three are two orders of magnitude cleaner than coal, oil, and natural gas.[1]

[1] 100–360 gC_{eq}/kW-h, with natural gas the lowest and coal the highest of the three.

Worldwide, if existing nuclear power plants were shut down and replaced with a mix of non-nuclear sources proportionate to what now exists, there would be an increase of 600 million tons of carbon emissions per year (IAEA, 2004)—equivalent to about twice the amount experts estimate will be avoided by adherence to the Kyoto Protocol in 2010.

Support for nuclear power and specific plans and actions to expand nuclear capacity in a number of countries are influencing global projections among nuclear insiders. The near-term projections released in 2004 by the International Energy Administration and the International Atomic Energy Agency were markedly higher than they were just four years ago. The most conservative projection predicted 427 gigawatts of global nuclear capacity in 2020, the equivalent of 127 more 1,000-megawatt plants than previous projections (ElBaradei, 2005).

Nuclear expansion is centered in Asia. Of the 25 reactors under construction worldwide, 17 are located either in China (including Taiwan), South Korea, North Korea, Japan, or India. Twenty of the last 30 reactors completed are in the Far East and south Asia (Langlois et al., 2005).

With 40 percent of the world population and the fastest growing economies in the world, the demand for new electric power in China and India is very high. The Chinese economy is expanding at a rate of 8 to 10 percent per year, and although China currently gets only 2.2 percent of its electricity from nuclear power, that percentage is scheduled to increase. By 2020, China plans to raise its total installed nuclear generating capacity from the current 6.5 gigawatts to 36 gigawatts, which will equate to 4 percent of total Chinese electricity supply (ElBaradei, 2005). India, which currently has nine plants under construction, plans to expand its nuclear capacity by a factor of 10 by 2022 and plans a 100-fold increase by mid-century. This sounds huge, but it works out to an average of about 9.2 percent per year, well below the pace of global nuclear capacity growth in the 1970s, which stood at 21 percent, but above the 1980s average of 8.7 percent (ElBaradei, 2005).

Although no U.S. plants have been ordered since the early 1970s, U.S. nuclear vendors have introduced technological innovations, such as advanced reactors, for certification by the U.S. Nuclear Regulatory Commission, which they have been marketing to other countries. The construction and operational experience of these vendors, and the experience of other multinational vendors, as well as countries developing indigenous designs, have kept nuclear technology moving forward.

The U.S.-led Generation IV International Nuclear Forum—a collegial effort by 10 countries—has published a road map for research and development on six innovative reactor concepts, such as the molten-salt reactor and the supercritical-water-cooled reactor. Innovative reactor and fuel-cycle technologies that address vulnerabilities related to safety, security, proliferation, and waste disposal and generate power at competitive prices are the most likely to be built. New nuclear plants will rely on passive safety features, fuel configurations with tighter control of sensitive nuclear materials, and design features that reduce construction times and lower operation and maintenance costs.

For nuclear energy to be considered a realistic solution to the energy needs of developing nations, a key feature will be size. Traditionally, nuclear plant designs have grown larger to take advantage of economies of scale. But smaller plants (less than 300 megawatts) that allow for more incremental investment, are more suited to smaller electrical-grid capacities and can be adapted more easily to other industrial applications, such as heating, seawater desalination, and the manufacture of chemical fuels.

A few of these designs are moving toward implementation. Russia has completed the design and licensing of a floating (barge-mounted) nuclear power plant, the KLT-40S, that takes advantage of the country's experience with nuclear-powered icebreakers and submarines. South Korea is making progress with a system-integrated, modular, advanced reactor (SMART). The Korean government plans to construct a one-fifth scale (65 megawatt) demonstration plant of this pressurized-water reactor by 2008, but has not yet announced a commercialization date for the full-scale (330 megawatt) plant. Among gas-cooled reactors, the South African pebble-bed modular reactor (PBMR), which features billiard-ball-sized self-contained fuel units and uses liquid sodium to transfer heat, is well under way. Preparation of the reactor site at Koeberg has begun, and fuel loading is anticipated for mid-2010. More innovative designs, still in development, employ modular cores that only require refueling every 30 years. This would address concerns about proliferation and reduce infrastructure needs.

A number of countries continue to reprocess spent nuclear fuel, and, in most cases, use it for the production of mixed-oxide fuel (MOX), which is then used as a reactor fuel for power generation. Because MOX requires plutonium, there are concerns about the proliferation potential of MOX, but France and Japan, among others, are proceeding nonetheless.

Transmutation, an idea that has been around for some time, is another approach to waste management. The basic goal is referred to as partitioning and transmutation—that is, trying to separate out the long-lived transuranic radionuclides (actinides, such as neptunium, americium, and curium, in particular), and using neutron bombardment in an accelerator-driven system (ADS) to burn up the nastiest bits of waste, making more electricity in the process. If these actinides could be converted into shorter-lived radionuclides, high-level radioactive waste would be much easier and less expensive to handle and dispose of. In addition to the actinides, longer-lived fission products, like technetium-99 and iodine-129, could also be burned up in an ADS.

HYDROGEN AND FUEL CELLS

Hydrogen has been much touted as an important fuel for the future. Hydrogen, the most abundant element, is used in liquid form to propel NASA space shuttles and other rockets and is the focus of national and international efforts to build the early stages of a hydrogen-based economy. Hydrogen fuel cells power the shuttle's electrical systems and emit water, which is reprocessed for the crew to drink. And, if hydrogen is produced by renewable processes, hydrogen-fueled applications do not create greenhouse gases.

However, hydrogen does not occur naturally as a gas; it occurs only in combination with other elements. Gaseous hydrogen to be used as a fuel can be made by separating it from hydrocarbons, usually generated by fossil fuels (or by electricity generated by fossil fuels) in a process called reformation. Hydrogen generated by this method is three to four times as expensive as gasoline as a transportation fuel, and because the hydrogen is generated from carbonaceous molecular systems, the reformation process still generates greenhouse gases. Hydrogen also can be created by combining zinc with water in the form of steam, which strips the oxygen from the water and leaves hydrogen. The challenge for industrializing this procedure, however, is finding an inexpensive way of turning the resulting zinc oxide back into metallic zinc so the material can be recycled (Economist, 2005).

Researchers at the Weizmann Institute of Science in Rehovot, Israel, have created a solar-power tower laboratory in which 64 mirrors track the sun, focusing its rays into a beam with a power of up to 300 kilowatts (Economist, 2005). The beam heats a mixture of zinc oxide and charcoal (pure carbon), which reacts with the oxygen in the zinc oxide and releases the zinc, which vaporizes and is then extracted and condensed into powder. The process does not produce greenhouse gases.

The powdered zinc can be used in zinc-air batteries which, though still experimental, might someday exceed the performance targets set by the U.S. Department of Energy for battery power and energy density in electric vehicles. With no moving parts or external tanks, the cell operates at room temperature and is simple to construct from readily available materials. The new zinc-air battery could be easily renewed at service stations and thus give electric vehicles the same driving range as gas-fueled vehicles, while eliminating exhaust pollution (Zyn Systems, 2005). But whether or not this method will reduce the cost of producing hydrogen for fuel has yet to be determined.

Researchers at Rensselaer Polytechnic Institute (RPI) are examining ways to develop materials for the next generation of fuel cells. The focus has been on hydrogen generation and storage, catalysis, electrochemistry, and polymer science. Another area of research is the application of nanomaterials in fuel cell and hydrogen research, including new materials to improve reliability, efficiency, and cost. New electrodes are also being developed, as are techniques for imaging an operating fuel cell.

Polymeric materials are central to proton-exchange membrane—or PEM—fuel cells. However, PEM fuel cells must be constantly hydrated, and maintaining the proper hydration level results in expensive and complex control schemes, which lead to reliability, cost, and robustness issues. In addition, PEM cells raise environmental issues, such as low-temperature operation.

Researchers at RPI have turned to a polymer called polybenzimidazole, or PBI. Currently, PBI fibers are used in protective apparel, such as turnout coats for firefighters and spacesuits for astronauts. PBI fibers have no melting point and are mildew resistant, age resistant, and abrasion resistant. At RPI a new generation of PBI has been developed, one that does not depend on the fiber process, does not require water for proton conductivity, and can operate at significantly higher temperatures than conventional fuel cells, thereby making it tolerant of impurities (e.g., carbon monoxide) in the fuel stream.

FUSION

Fusion has long captured the imagination as a source of virtually unlimited energy—if it could be contained and controlled. Efforts to create nuclear fusion using strong magnetic fields or large lasers to contain the plasma in which the fusion occurs have so far failed to produce more energy than they use.

One Rensselaer researcher is looking into sonofusion, a new form of nuclear fusion. In sonofusion, deuterated acetone, in which hydrogen is substituted with deuterium, is placed in a flask. The rapid contractions and expansions of a piezoelectric ceramic ring on the outside of the flask send pressure waves through the liquid. At points of low pressure, the liquid is bombarded with neutrons, creating clusters of bubbles, which greatly expand during the low-pressure conditions, and then, as the pressure begins to increase again, implode, sending shock waves toward the center of the bubbles. This creates very high pressures and temperatures in an extremely small region of the collapsing bubbles. Careful measurements show that deuterium atoms located there have fused, releasing additional neutrons into the liquid and creating tritium.

When the research team first announced successful sonofusion in 2002 in the journal *Science*, their paper was met with skepticism (Taleyarkhan et al., 2002). But two years later, the team announced that it had successfully duplicated its results using more sensitive instrumentation (Taleyarkhan et al., 2004). At least five other research groups are now trying to reproduce the results, and one recently announced independent confirmation (Xu and Butt, 2005).

Although the amount of energy being produced with sonofusion is extremely small, researchers hope that the process can be scaled up successfully. A recently formed consortium—the Acoustic Fusion Technology Energy Consortium (AFTEC)—is exploring the potential of sonofusion. Members of the consortium include Boston University, UCLA, the University of Mississippi, the University of Washington, Purdue University, the Russian Academy of Sciences, and Rensselaer.

Although much more research is needed, if sonofusion reactors ever are able to produce usable quantities of power, the process might become a major energy source that operates without producing the radioactive waste produced by nuclear fission reactors. Fusion power will require methods of scaling-up the process and making it self-sustaining.

THE "QUIET CRISIS"

The reality is that we can no longer just drill our way to global energy security. We must innovate our way to energy security—we must find new technologies that uncover new fossil energy sources, that conserve energy, that protect the environment, and that provide multiple, sustainable sources of energy. It is clear that the technological developments I have outlined are a long way from being viable energy "solutions." But, innovation itself is a kind of energy that multiplies insights and advances discovery, enabling the best minds to work—in concert—on what may appear to be overwhelming challenges.

Nurturing our human capacity for innovation requires a configuration of elements in which multidisciplinarity and interdisciplinarity, and cooperation and collaboration interface. Indeed, this is the future of engineering. It is the future of science. It is the future of discovery and innovation. And, this concept is at the core of this symposium. With achievements to your credit and time for future achievements, you are the present and future of engineering.

But, who will come after you? Engineering of the future requires people, and we are no longer turning out a sufficient number of people to replace the ones we have now. Enrollment of American students in physical sciences, mathematics, and engineering has declined severely over the past decade. Unless and until we have cohorts of young people who are ready to step into the laboratories and design studios to replace the scientists and engineers who will soon be retiring in great numbers, we will not have the capacity for the kind of innovation we need. At the same time, as other nations invest in their own education and research enterprises, and as globalization offers employment for their scientists and engineers at home or elsewhere, the flow to the United States of talented international scientists and engineers and graduate students is slowing.

The net result is that the American innovation enterprise, which has fueled and sustained our economic growth, our standard of living, and our security and has made us a global leader, soon may lack the critical mass of scientists and engineers necessary for the next innovations upon which new industries will be built and upon which solutions to global challenges depend. I call this urgent situation the "quiet crisis," because the forces have come together over a period of time, with little notice, and have accelerated recently.

We need young people whose curiosity has been whetted, whose imaginations have been sparked, whose eagerness for science and mathematics has been awakened, and who are ready to be nurtured as they pass through the engineering and science pipeline. Where will they come from? The demographics in this country have shifted dramatically over the last couple of decades. A new majority now comprises young women and the racial and ethnic groups that traditionally have been underrepresented in engineering and science. These young people—even the brightest among them—often are not encouraged to pursue preparatory coursework that would enable them to pursue an engineering or science degree at an advanced level—even though their enrollment in higher education is increasing faster than the enrollment of traditional engineering and science students. In addition, we do not yet have faculty and upperclassmen from the new majority who can serve as role models and mentors to shepherd these nontraditional students.

And yet, if we are to build a future cohort of engineers and scientists, this is where we must look. One of the major challenges to our entire education system—K–12 and higher education—is to reach out to these students, the underrepresented majority in science, engineering, and technology, and help them find their way in and help them stay in for the duration. This must be done against the backdrop of encouraging all of our young people to take on the challenges of science and advanced mathematics in primary and secondary school and to consider engineering, science, and related majors in college and beyond.

The "quiet crisis" is finally being noticed. In the four or five years since I have been working on and speaking to this issue, a growing understanding and concern has developed—in every sector—that the issue is real and must be addressed—and quickly. More and more sectors and organizations are asking that attention be paid to the matter and that practices be changed to address it.

CONCLUSION

The need to address the long-term effects of Hurricane Katrina could very well increase awareness of this concern, and one would hope that the need for new energy sources and resources will inspire the next generation of engineers and scientists in the same way the Soviet launch of Sputnik inspired the young people of my generation. It should be clear that a sustainable global energy framework capable of meeting the energy needs of citizens without causing irreparable environmental damage will require continuing technological advances that modify our current production and uses of energy.

The challenges of energy security and the challenges of other threats without borders that beset our young century and are disrupting global security require new strategies, new alternatives, new approaches, and new ways of thinking. Every profession will be challenged to find new ways to work and think, to plan and collaborate, to innovate and discover.

This challenge might be called a "hidden benefit"—a silver lining, perhaps. The challenge is forcing us to move into a higher sphere where more is at risk, setting the bar higher for humankind. I believe humankind will rise to the challenge. I believe you will rise to the challenge.

Thank you.

REFERENCES

Economist. 2005. Solid Sunshine. Economist 376(8439): 68.

EIA (Energy Information Administration). 2005. International Energy Outlook 2005. Washington, D.C.: U.S. Department of Energy. Available online at: *http://www.eia.doe.gov/oiaf/ieo/*.

ElBaradei, M. 2005. Nuclear Power: Preparing for the Future. Statements of the General Director at the International Conference on Nuclear Power for the 21st Century, March 21, 2005, Paris, France. Available online at: *http://www.iaea.org/NewsCenter/Statements/2005/ebsp2005n004.html*.

IAEA (International Atomic Energy Association). 2004. Energising the Future: The Power of Innovation. June 2004, p. 6. Available online at: *http://www.iaea.org/Publications/Magazines/Bulletin/Bull461/index.html*.

IEA (International Energy Agency). 2004. World Energy Outlook 2004. Paris, France: IEA. Available online at: *http://www.iea.org/bookshop/add.aspx?id=180*.

Langlois, L.M., A. McDonald, H.-H. Rogner, and F.L. Toth. 2005. Nuclear Power: Rising Expectations. Pp. 15–20 in International Power and Utilities Finance Review 2005/2006. London, U.K.: Euromoney Institutional Investor PLC. Available online at: *www.iaea.org/OurWork/ST/NE/Pess/assets/NP_RisingExpect.pdf*.

Lucchetti, A., V. Bauerlein, and K. Hudson. Storm Leaves Gulf Coast Devastated. *Wall Street Journal*, August 31, 2005, p. A.1.

Osborne, D. 2005. Billion People Face Water Shortages, Warns UN. *New Zealand Herald*, March 23, 2005.

Taleyarkhan, R.P., C.D. West, J.S. Cho, R.T. Lahey Jr., R.I. Nigmatulin, and R.C. Block. 2002. Evidence for nuclear emissions during acoustic cavitation. Science 295(5561): 1868–1873.

Taleyarkhan, R.P., J.S. Cho, C.D. West, R.T. Lahey Jr., R.I. Nigmatulin, and R.C. Block. 2004. Additional evidence for nuclear emissions during acoustic cavitation. Physical Review E 69(3): 036109.

Xu, Y., and A. Butt. 2005. Confirmatory experiments for nuclear emissions during acoustic cavitation. Nuclear Engineering and Design 235(10-12): 1317–1324.

Zyn Systems. 2005. Technology Commercialization Opportunity: Zinc-Air Battery. Available online at: *http://www.zyn.com/flcfw/fwtproj/ZincAirB.htm*.

Appendixes

Contributors

Stuart B. Adler is an assistant professor in the Department of Chemical Engineering at the University of Washington. His areas of research include solid-state electrochemical engineering, electrocatalysis, ionic transport, ceramics, and fuel cells. Dr. Adler received a B.S. from the University of Michigan and a Ph.D. from the University of California, Berkeley.

Peter N. Belhumeur, a professor in the Department of Computer Science at Columbia University and director of the Laboratory for the Study of Visual Appearance, received an Sc.B. in information sciences from Brown University in 1985 and a Ph.D. in engineering sciences from Harvard University in 1993. In 1994, he was a postdoctoral fellow at the University of Cambridge Isaac Newton Institute for Mathematical Sciences. After holding several positions in electrical engineering at Yale University, he joined the faculty of Columbia University as a professor of computer science in 2002. Dr. Belhumeur's research on illumination, reflectance, and shape and their relation to visual appearance is concentrated on the representation and recognition of objects under variable illumination and the estimation of the geometry of objects from low-level cues, such as image brightness, binocular stereopsis, and motion. He is a recipient of the Presidential Early Career Award for Scientists and Engineers (PECASE), the National Science Foundation Career Award, the Siemens Best Paper Award at the 1996 IEEE Conference on Computer Vision and Pattern Recognition, and the Olympus Prize at the 1998 European Conference of Computer Vision. His research is funded by the National Science Foundation, the National Institutes of Health, the Army Research Office, and the Defense Advanced Research Projects Agency.

Shirley Ann Jackson is the 18th president of Rensselaer Polytechnic Institute, Troy, New York, and Hartford, Connecticut, the oldest technological research university in the United States. Dr. Jackson has held senior leadership positions in government, industry, research, and academia. She is immediate past president of the American Association for the Advancement of Science (AAAS) and current chair of the AAAS Board of Directors; a member of the National Academy of Engineering; and a Fellow of the American Academy of Arts and Sciences and the American Physical Society. She also holds advisory positions and is an active participant in other prestigious national organizations: trustee of the Brookings Institution; life member of the Massachusetts Institute of Technology (MIT) Corporation; member of the Council on Foreign Relations; member of the Executive Committee of the Council on Competitiveness; board member of Georgetown University and Rockefeller University; member of the Board of Directors of the New York Stock Exchange and the Board of Regents of the Smithsonian Institution; and director of several major corporations. In 1995, President William J. Clinton appointed Dr. Jackson chair of the U.S. Nuclear Regulatory Commission (NRC). During her tenure at the NRC (1995–1999), she reorganized the agency and revamped its regulatory approach by moving strongly toward risk-informed, performance-based regulation. Prior to that, Dr. Jackson was a theoretical physicist at the former AT&T Bell Laboratories and a professor of theoretical physics at Rutgers University. She holds an S.B. in physics and Ph.D. in theoretical elementary particle physics from MIT and 31 honorary doctoral degrees.

Daniel M. Kammen, Class of 1935 Distinguished Professor of Energy, holds appointments in the Energy and Resources Group, Department of Nuclear Engineering, and Goldman School of Public Policy at the University of California, Berkeley. He is the founding director of the Renewable and Appropriate Energy Laboratory and co-director of the Berkeley Institute of the Environment. Until 1999, Dr. Kammen was on the faculty of the Woodrow Wilson School of Public and International Affairs at Princeton University, where he taught courses in science and technology analysis and policy. He is currently involved in long-term research projects on energy and development in Africa and Latin America and laboratory and modeling projects on clean energy futures. Dr. Kammen is the author of a book on environmental, technological, and health risks, *Should We Risk It?* (Princeton University Press, 2001), more than 160 journal articles, and numerous reports on renewable energy and development. He has been featured on radio, network, and public broadcasting programs and in print as an analyst of energy, environmental, and risk policy issues and current events. Dr.

Kammen received an A.B. in physics from Cornell University and an M.A. and Ph.D. in physics from Harvard University.

Jay D. Keasling, a professor in the Departments of Chemical Engineering and Bioengineering at the University of California, Berkeley, is also a faculty scientist and director of the Physical Biosciences Division at the Lawrence Berkeley National Laboratory and director of the Berkeley Center for Synthetic Biology. Dr. Keasling, a pioneer in the field of synthetic biology, conducts research on the engineering of microorganisms for the environmentally friendly synthesis of small molecules and the degradation of environmental contaminants. His laboratory has engineered *Escherichia coli* to produce polymers; a precursor to the anti-malarial drug artemisinin; and *Pseudomonas putida* to accumulate uranium and degrade nerve agents. Dr. Keasling received his B.S. in chemistry and biology from the University of Nebraska in 1986 and his Ph. D. in chemical engineering from the University of Michigan in 1991. From 1991 to 1992, he was a postdoctoral student in biochemistry at Stanford University.

Natalia L. Komarova is an assistant professor in the Department of Mathematics at the University of California, Irvine. Previously, she was assistant professor of mathematics at Rutgers University and a member of the School of Mathematics and the Center for Systems Biology at the Institute for Advanced Study, Princeton. Her main interest is in the interface between mathematics and life sciences; she uses mathematics to address questions posed by biology (in the field of cancer and viruses), history, and linguistics. In 2002, Student Achievement & Advocacy Services awarded her the first Prize for Promise, given to a "woman of exceptional ability, ambition, brilliance, courage, dedication and vision." In 2005, she received the Alfred P. Sloan Fellowship in mathematics. Dr. Komarova is coauthor (with D. Wodarz) of *Computational Biology of Cancer: Lecture Notes and Mathematical Modeling* (World Scientific, 2005). She received an M.S. in physics from Moscow State University and an M.S. and Ph.D. (1998) in applied mathematics from the University of Arizona.

Kurt L. Kornbluth received a B.S. from Michigan State University and an M.S. from San Francisco State University in mechanical engineering and is currently a Ph.D. student in mechanical engineering at the University of California, Davis. Previously, as an instructor at the Edgerton Center at the Massachusetts Institute of Technology, he supervised student implementation of international development projects. As part of his Ph.D. research, Mr. Kornbluth worked as a design engineer at DEKA Research and Development in Manchester, New Hampshire, on the Stirling Engine Village Power Project, for which he developed a biogas venturi burner for use in developing countries. Previous to that he was director of operations and development engineer at Whirlwind Wheelchair International in San Francisco, where he managed international capacity-building projects, implemented technical aspects, including wheelchair design and production tooling, and trained technicians in developing countries. Mr. Kornbluth designed and produced a prototype of a high-stability, compact, omni wheelchair, for which he was honored by the Rehabilitation Engineering Society of North America Paralyzed Veterans of America. As a participant in Whirlwind Africa 1, he worked with a design team to establish design criteria for a wheelchair for small-scale production in developing countries and designed a technology transfer kit for distributing the Africa-1 wheelchair design.

Daniele S. Lantagne is a staff engineer at the Centers for Disease Control and Prevention (CDC), where her primary role is to provide technical assistance and program support for projects to ensure safe water systems in developing countries. Her responsibilities include working with project partners in, and traveling to, developing countries to implement, provide technical assistance and program support, and assess water systems and other point-of-use water treatment projects; answering technical questions, writing informational fact sheets, and maintaining the safe water system website; and providing engineering expertise to the Foodborne and Diarrheal Disease Branch of CDC. Prior positions include principal, Alethia Environmental; lecturer, Massachusetts Institute of Technology (MIT); program director, Ipswich River Watershed Association; outreach program coordinator, MIT Edgerton Center; and environmental engineer, Louis Berger & Associates Inc. Ms. Lantagne is a member of Sigma Xi and was a recipient of the Ipswich River Watershed Association Leadership Award. She received a B.S. and M.S. in environmental engineering from MIT.

Michael D. McGehee is an assistant professor of materials science and engineering at Stanford University. His research is focused on the electrical and optical properties of organic semiconductors, the self-assembly of inorganic nanostructures with organic structure-directing agents, and the fabrication of low-cost organic photovoltaic cells, light-emitting diodes, and transistors. Dr. McGehee is the recipient of a Henry and Camille Dreyfus New Faculty Award, a National Science Foundation CAREER Award, and the Dupont Young Professor Award. He received an A.B. in physics from Princeton University and a Ph.D. in materials science from the University of California, Santa Barbara.

Matthai Philipose, a researcher at the Intel Research Laboratory in Seattle, conducts research on statistical reasoning and programming languages, with a special focus on systems that can automatically recognize human activities and the application of these systems to caregiving. Dr. Philipose heads the System for Human Activity Recognition and Prediction (SHARP) Project, which is working on the development of sensor-based systems that can recognize a large number, perhaps thousands, of day-to-day activities. He received a B.S. in computer science from Cornell University and an M.S. and a Ph.D. in computer science from the University of Washington.

P. Jonathon Phillips is program manager for the Face Recognition Grand Challenge and Iris Challenge Evaluation at the National Institute of Standards and Technology (NIST), as well as test director for the Face Recognition Vendor Test (FRVT) 2005. From 2000 to 2004, Dr. Phillips was assigned to the Defense Advanced Research Projects Agency as program manager for the Human Identification at a Distance Program. He received a U.S. Department of Commerce Gold Medal for his work as a test director for FRVT 2002. His current research interests include computer vision, face recognition, biometrics, digital video processing, methods of evaluating biometric algorithms, and computational psychophysics. Prior to joining NIST, Dr. Phillips developed and designed the FERET database collection and FERET evaluations at the U.S. Army Research Laboratory. He has organized three conferences and workshops on face recognition and three on empirical evaluation and coedited three books on face recognition and empirical evaluation. Dr. Phillips has been guest editor of special issues or sections of the *IEEE Transactions on Pattern Analysis and Machine Intelligence* and *Computer Vision and Image Understanding*. He is an associate editor for *IEEE Transactions on Pattern Analysis and Machine Intelligence* and guest editor of a special issue of *Proceedings of the IEEE* on biometrics. He is also a member of the IEEE. Dr. Phillips received a B.S. in mathematics and an M.S. in electronic and computer engineering from George Mason University and a Ph.D. in operations research from Rutgers University.

Sunita Satyapal, team leader for hydrogen storage at the U.S. Department of Energy (DOE), oversees the National Hydrogen Storage Project, which has a $150 million budget for five years. As part of President Bush's Hydrogen Fuel Initiative, the focus of this project is on the development of materials-based technologies, such as metal hydrides, chemical hydrides, or high-surface-area sorbents, to store hydrogen for hydrogen-powered vehicles. Dr. Satyapal joined DOE in 2003 after working for eight years in industry at United Technologies Research Center and UTC Fuel Cells. During those eight years, she was responsible for managing various research groups of 15 to 50 scientists, engineers, and technicians working on a broad range of chemistry and chemical engineering technologies. She has also conducted research on laser diagnostics for the combustion of chemical warfare agents in the Department of Applied and Engineering Physics at Cornell University and taught chemistry at Vassar College. Dr. Satyapal is the author of numerous technical publications related to chemistry/chemical engineering and the owner of 10 patents. She received a B.A. from Bryn Mawr College and a Ph.D. from Columbia University.

Zoltán Toroczkai is deputy director of the Center for Nonlinear Studies at Los Alamos National Laboratory (LANL). From 2000 to 2002, he held a Director's Fellowship at LANL, and from 1998 to 2000, he was a research associate at the University of Maryland, College Park. His current research is focused on complex networks, specifically statistical physics of complex networks with applications to infrastructure networks and social systems; agent-based systems modeling, multiplayer games, game theory, collective intelligence, and optimization; massively parallel computation; and statistical physics and nonequilibrium statistical mechanics, nonlinear dynamical systems and chaos, population dynamics, and species coexistence. Dr. Toroczkai is the author or coauthor of more than 50 publications in peer-reviewed journals. He received a master's degree from Babes-Bolyai University in Romania and a Ph.D. in theoretical physics from Virginia Polytechnic and State University in 1997.

Alessandro Vespignani, professor of informatics at Indiana University, earned his Ph.D. from the University of Rome La Sapienza. After holding research positions at Yale University and Leiden University, he joined a group doing research on condensed matter at the International Center for Theoretical Physics (UNESCO) in Trieste, where he headed research and teaching activities for more than five years. He then joined the French National Council for Scientific Research, where he continued his academic work at the Laboratoire de Physique Theorique of the University of Paris-Sud. His recent research has been focused on the interdisciplinary application of statistical physics and numerical simulation methods in the analysis of epidemic and spreading phenomena and the study of biological, social, and technological networks. Dr. Vespignani is the author of more than 100 scientific papers on the properties and characterization of nonequilibrium phenomena, critical phase transitions, and complex systems and coauthor (with R. Pastor-Satorras) of *Evolution and Structure of the Internet* (Cambridge University Press, 2004). Dr. Vespignani was one of five scientists nominated for the Wired Magazine Rave Award in science for 2004.

Julie Beth Zimmerman, an assistant professor in the Department of Civil Engineering at the University of Virginia, teaches and conducts research on pollution prevention, green engineering, green chemistry, and sustainability. As engineer/program coordinator for the Environmental Protection Agency (EPA) Office of Research and Development National Center for Environmental Research, she manages academic grants for the Technologies for Sustainable Environment research program; coordinates Small Business Innovation Research contracts for clean technologies, pollution prevention, and research on waste minimization; initiated the P3 Award, a national student design competition; and initiated benchmarking of the integration of sustainability into engineering curricula at U.S. institutions of higher education. Dr. Zimmerman is a member of the steering committee for the U.S. Partnership for the U.N. Decade for

Education for Sustainable Development and has served on programming committees for the International Green Chemistry and Engineering Conference and the EPA Annual Green Chemistry and Engineering Conference. She is a member of numerous professional associations, including the American Chemical Society, American Society of Civil Engineers, American Society for Engineering Education, and Engineers without Borders. Dr. Zimmerman received a B.S. from the University of Virginia and an M.S. and interdepartmental Ph.D. from the College of Engineering and School of Natural Resources and Environment, University of Michigan.

Program

NATIONAL ACADEMY OF ENGINEERING

2005 U.S. Frontiers of Engineering Symposium
September 22–24, 2005

Chair: Pablo G. Debenedetti, Princeton University

ID AND VERIFICATION TECHNOLOGIES

Organizers: Stephen S. Intille, Massachusetts Institute of Technology, and
Visvanathan Ramesh, Siemens Corporate Research, Inc.

Ongoing Challenges in Face Recognition
Peter N. Belhumeur, Columbia University

Designing Biometrict Evaluations and Challenge Problems for Face Recognition Systems
P. Jonathon Phillips, National Institute of Standards and Technology

Large-Scale Human Activity Recognition Using Ultra-Dense Sensing
Matthai Philipose, Intel Research

***

ENGINEERING FOR DEVELOPING COMMUNITIES

Organizers: Garrick E. Louis, University of Virginia, and
Amy Smith, Massachusetts Institute of Technology

Challenges in Implementation of Appropriate Technology Projects: The Case of the DISACARE Wheelchair Center in Zambia
Kurt Kornbluth, University of California, Davis

Engineering Inputs to Increase Impact of the CDC Safe Water System Program
Daniele S. Lantagne, U.S. Centers for Disease Control and Prevention

Sustainable Development Through the Principles of Green Engineering
Julie Beth Zimmerman, University of Virginia and
U.S. Environmental Protection Agency

"We Don't Tenure Mother Teresa":
Evolving Science and Engineering Research to Value the Planet
Daniel M. Kammen, University of California, Berkeley

***

Dinner Speech—*Engineering for a New World*
Dr. Shirley Ann Jackson, President, Rensselaer Polytechnic Institute

***

ENGINEERING COMPLEX SYSTEMS
Organizers: Luis A. Nunes Amaral, Northwestern University, and
Kelvin H. Lee, Cornell University

Complex Networks: Ubiquity, Importance, and Implications
Alessandro Vespignani, Indiana University

Engineering Bacteria for Drug Production
Jay D. Keasling, University of California, Berkeley and
Lawrence Berkeley National Laboratory

Population Dyanimcs of Human Language: A Complex System
Natalia L. Komarova, University of California, Irvine

Agent-based Modeling as a Decision Making Tool:
How to Halt a Smallpox Epidemic
Zoltán Toroczkai, Los Alamos National Laboratory

***

ENERGY RESOURCES FOR THE FUTURE

Organizers: Allan J. Connolly, GE Energy, and
John M. Vohs, University of Pennsylvania

Future Energy
John K. Reinker, GE Global Research

Organic-based Solar Cells
Michael D. McGehee, Stanford University

Hydrogen Production and Storage Research and Development Activities at the U.S. Department of Energy
Sunita Satyapal, U.S. Department of Energy

Fuel Cells: Current Status and Future Challenges
Stuart B. Adler, University of Washington

Participants

NATIONAL ACADEMY OF ENGINEERING

2005 U.S. Frontiers of Engineering Symposium
September 22–24, 2005

Stuart Adler
Assistant Professor
Department of Chemical Engineering
University of Washington

Apoorv Agarwal
EVA Engine PMT Leader
Ford Motor Company

Carina Maria Alles
Engineering Consultant
DuPont

Luis A. Nunes Amaral
Associate Professor
Department of Chemical and Biological Engineering
Northwestern University

Adjo Amekudzi
Associate Professor
School of Civil and Environmental Engineering
Georgia Institute of Technology

Alyssa Apsel
Clare Booth Luce Assistant Professor
School of Electrical Engineering
Cornell University

Treena Arinzeh
Assistant Professor
Department of Biomedical Engineering
New Jersey Institute of Technology

Patrick Barge
Executive Director of Engineering
Cummins Inc./Fleetguard

Peter Belhumeur
Professor
Department of Computer Science
Columbia University

Serge Belongie
Assistant Professor
Department of Computer Science and Engineering
University of California, San Diego

John Bettler
Senior Engineer
Com Edison

Sujata Bhatia
Medical Research Scientist
DuPont Central Research & Development

Brian Blake
Associate Professor
Department of Computer Science
Georgetown University

Tracy Camp
Associate Professor
Department of Math and Computer Sciences
Colorado School of Mines

Kurt Casby
Capacitor Development Engineering Manager
Energy and Components Center
Medtronic

Sila Çetinkaya
Associate Professor
Department of Industrial Engineering
Texas A&M University

Jason Chen
Project Leader/Senior Staff Engineer
Alcoa Technical Center

Trishul Chilimbi
Researcher
Runtime Analysis and Design
Microsoft Research

Allan Connolly
General Manager
Power Generation Systems Engineering
GE Energy

Robert Crane
Section Head, Catalyst Manufacturing
ExxonMobil Research & Engineering

Scott Craver
Assistant Professor
Department of Electrical and Computer Engineering
Binghamton University

Mary Crawford
Principal Member of Technical Staff
Sandia National Laboratories

Kristin Culler
Flight Controls Analysis Engineer
Boeing Space & Communications
Boeing Company

Juan de Bedout
Manager, Electric Power and Propulsion Systems Lab
GE Global Research

Pablo Debenedetti
Class of 1950 Professor
Department of Chemical Engineering
Princeton University

Kristian Debus
Senior Engineering Specialist
Bechtel National, Inc.

Michael Deem
John W. Cox Professor of Biochemical and Genetic Engineering and Professor of Physics and Astronomy
Rice University

Marina Despotopoulou
Senior Research Engineer
Arkema, Inc.

Ram Devanathan
Senior Research Scientist
Fundamental Science Doctorate
Pacific Northwest National Laboratory

David Doman
Senior Aerospace Engineer
Air Force Research Laboratory

Alexander Driskill-Smith
Advisory Engineer/Scientist
Hitachi Global Storage Technologies

Marija Drndic
Assistant Professor
Department of Physics and Astronomy
University of Pennsylvania

Richard Elander
Senior Engineer II
National Renewable Energy Laboratory

Bogdan Epureanu
Assistant Professor
Department of Mechanical Engineering
University of Michigan

Ireena Erteza
Principal Member of the Technical Staff
Sandia National Laboratories

Silvia Ferrari
Assistant Professor
Department of Mechanical Engineering and Materials Science
Duke University

Gary Fogel
Vice President
Natural Selection, Inc.

Russell Ford
Global Technology Leader
Water Treatment
CH2M Hill

Robert Franceschini
Chief Systems Engineer
Science Applications International Corp.

Juan Gilbert
Assistant Professor
Department of Computer Science and Software Engineering
Auburn University

William Grieco
Program Manager
Emerging Technologies
Rohm and Haas Company

Arun Hampapur
Manager, Exploratory Computer Vision Group
T.J. Watson Research Center
IBM Corporation

Glenn Hart
Software Engineer
Westinghouse Electric Company LLC

William Hartt
Technology Leader, Complex Fluid Dynamics
Procter & Gamble

Youssef Hashash
Associate Professor
Department of Civil Engineering
University of Illinois, Urbana-Champaign

Michael Helmbrecht
President and Chief Executive Officer
Iris AO, Inc.

Mark Hersam
Assistant Professor
Department of Materials Science and Engineering
Northwestern University

Cynthia Hoover
Director, Electronics and Analytical R&D
Praxair, Inc.

Scott Humphreys
RFMD Fellow
RF Micro Devices, Inc.

Stephen Intille *(unable to attend)*
Technology Director, House_n Consortium
Department of Architecture
Massachusetts Institute of Technology

Christopher Jones
Associate Professor
School of Chemical and Biomolecular Engineering
Georgia Institute of Technology

Daniel Kammen
Professor in the Energy and Resources Group and Professor of Public Policy, and Director, Renewable and Appropriate Energy Laboratory
Department of Nuclear Engineering
University of California, Berkeley

Jay Keasling
Professor
Department of Chemical Engineering
University of California, Berkley

Maryam Khanbaghi
Senior Research Engineer, Controls
Corning, Inc.

Erica Klampfl
Technical Expert
Infotronics and Systems Analytics Department
Ford Motor Company

Melissa Knothe Tate
Associate Professor, Director of the Think Tank for Multiscale Computational Modeling of Biomedical and Bio-inspired Systems
Departments of Mechanical and Aerospace and Biomedical Engineering
Case Western Reserve University

Kurt Kornbluth
Department of Mechanical Engineering
University of California, Davis

David Kosson
Chair and Professor of Civil and Environmental Engineering, Professor of Chemical Engineering
Department of Civil and Environmental Engineering
Vanderbilt University

Ananth Krishnamurthy
Assistant Professor
Department of Decision Sciences and Engineering Systems
Rensselaer Polytechnic Institute

Vikram Kumar
President and Chief Executive Officer
Dimagi, Inc.

Christine Lambert
Technical Expert
Scientific Research Laboratory
Ford Motor Company

Daniele Lantagne
Safe Water System Staff Engineer
Foodborne and Diarrheal Disease Branch
Centers for Disease Control and Prevention

Kelvin Lee
Associate Professor
School of Chemical and Biomolecular Engineering
Cornell University

Hod Lipson
Assistant Professor
School of Mechanical and Aerospace Engineering
Cornell University

Jie Liu
Researcher
Networked Embedded Computing Group
Microsoft Research

Zhu Liu
Technical Specialist
AT&T Labs Research

Garrick Louis
Associate Professor
Department of Systems and Information Engineering
University of Virginia

Hiroshi Matsui
Associate Professor
Department of Chemistry
City University of New York, Hunter College

Cary McConlogue
Senior Research Investigator
Pharmaceutical Research Institute
Bristol-Myers Squibb Company

Michael McGehee
Assistant Professor
Department of Materials Science and Engineering
Stanford University

Scott Miller
Chemical Engineer
GE Global Research

Michael Occhionero
Senior Advisory Engineer
National and Homeland Security
Idaho National Laboratory

Marcia O'Malley
Assistant Professor
Department of Mechanical Engineering and Materials Science
Rice University

Jonathan Owen
Staff Research Engineer
General Motors Research and Development Center

Janet Pan
Associate Professor of Electrical Engineering and of Applied Physics
Department of Electrical Engineering
Yale University

Pamela Patterson
Research Staff Member
HRL Laboratories, LLC

Sameer Pendharkar
Distinguished Member of Technical Staff
Device Engineering, Mixed Signal Technology Development
Texas Instruments, Inc.

Matthai Philipose
Researcher
Intel Research Laboratory

Julia Phillips
Director
Physical, Chemical, and Nano Sciences Center
Sandia National Laboratories

P. Jonathon Phillips
Program Manager
Information Access Division-Image Group
National Institute of Standards and Technology

Tresa Pollock
Professor
Department of Materials Science and Engineering
University of Michigan

Vivek Prabhu
Chemical Engineer
National Institute of Standards and Technology

Visvanathan Ramesh
Department Head
Siemens Corporate Research, Inc.

John Reinker
Leader, Energy and Propulsion Technologies
GE Global Research

George Rittenhouse
Vice President, Wireless Research
Bell Laboratories, Lucent Technologies

Rodrigo Salgado
Professor
School of Civil Engineering
Purdue University

Sunita Satyapal
Hydrogen Storage Team Leader
Office of Energy Efficiency and Renewable Energy
U.S. Department of Energy

William Schneider
Associate Professor
Department of Chemical and Biomolecular Engineering, Concurrent in Chemistry
University of Notre Dame

Cyrus Shahabi
Associate Professor
Department of Computer Science
University of Southern California

Sandeep Shukla
Assistant Professor
Department of Electrical and Computer Engineering
Virginia Polytechnic Institute and State University

Stanislav Shvartsman
Assistant Professor
Department of Chemical Engineering
Princeton University

Dan Sievenpiper
Senior Research Staff Engineer
HRL Laboratories, LLC

Kiruba Sivasubramaniam
Lead Technologist, Multimegawatt Electric Power Generator
Electromagnetics and Superconductivity
GE Global Research

Amy Smith
Instructor
Edgerton Center
Massachusetts Institute of Technology

Christina Smolke
Assistant Professor
Department of Chemical Engineering
California Institute of Technology

James Sowder
Chief Technologist and Technical Fellow
Northrop Grumman Information Technology

Abraham Stroock
Assistant Professor
School of Chemical and Biomolecular Engineering
Cornell University

Lester Su
Assistant Professor
Department of Mechanical Engineering
Johns Hopkins University

Julie Swann
Assistant Professor
School of Industrial and Systems Engineering
Georgia Institute of Technology

Aaron Thode
Research Scientist, Undersea Acoustics
Scripps Institution of Oceanography

Zoltán Toroczkai
Deputy Director
Center for Nonlinear Studies
Los Alamos National Laboratory

Thomas Truskett
Assistant Professor
Department of Chemical Engineering
University of Texas, Austin

Uday Turaga
Associate Scientist
ConocoPhillips Company

Jeffrey Varner
Professor
School of Chemical and Biomolecular Engineering
Cornell University

Alessandro Vespignani
Professor
School of Informatics
Indiana University

John Vohs
Carl V. Patterson Professor and Chair, Department of Chemical and Biomolecular Engineering
University of Pennsylvania

Cliff Wang
Program Director
U.S. Army Research Office

Min Wu
Assistant Professor
Department of Electrical and Computer Engineering
University of Maryland, College Park

Sophia Zamor
Senior Cross Sourcing Specialist
Cartago Operations
Kimberly-Clark Corporation

Julie Zimmerman
Assistant Professor
Department of Civil Engineering
University of Virginia
Engineer/Program Coordinator
National Center for Environmental Research
U.S. Environmental Protection Agency

Guests

Daniel Berg
Institute Professor of Science and Technology
Rensselaer Polytechnic Institute

Richard Buckius
Interim Assistant Director for Engineering Directorate
National Science Foundation

Eduardo Misawa
Program Director, Dynamical Systems
Intelligent Civil and Mechanical Systems
National Science Foundation

Sarah Tegan
Editorial Associate
Proceedings of the National Academy of Sciences

James Tien
Yamada Corporation Professor
Rensselaer Polytechnic Institute

Dinner Speaker

Shirley Ann Jackson
President
Rensselaer Polytechnic Institute

GE Global Research

Sean Connolly
Event Coordinator
GE Global Research Center

Ingrid Cullen
Technology/Marketing Executive Admin
GE Plastics

National Academy of Engineering

Wm. A. Wulf
President

Lance A. Davis
Executive Officer

Janet Hunziker
Program Officer

Gin Bacon
Senior Program Assistant